Walking Each Other Home

HEART-HEALING ENCOURAGEMENT BY
MELANIE BOSTER

WITH STUNNING PHOTOGRAPHY BY
DAVID W. BOSTER

WHITESTONE PUBLISHING
STONEWOOD, WV, USA

Walking Each Other Home
By Melanie Boster

Photography by David W. Boster

ISBN-10: 1628830123

ISBN-13: 978-1628830125

Library of Congress Control Number: 2016907627

Copyright © 2016 Melanie Boster
Patriot, OH 45658

All rights reserved. No portion of this publication may be reproduced, stored in an electronic system or transmitted in any form or by any means, electronic, mechanical, photocopy, recording, or otherwise, without the prior permission of the author.
Brief quotations may be used in literary reviews.

Printed in United States of America.

This book may be purchased at:
http://amazon.com
http://bosterbooks.com
or wherever books are sold.

CHRISTIAN RESOURCES | INSPIRATIONAL NOVELS | CHILDREN'S BOOKS

Table of Contents

The Journey Begins 1
Walking The Rails 4
Fasten Your Seatbelts 7
Some Things You Never Forget 10
A Little Rearview Perspective 13
Traveling Light 16
Dawn 19
Ready to Run 21
Adventure Time 24
Cool Running 27
Mud 29
What's That Smell? 33
I Yam What I Yam 37
A Reason To Be 40
Too Good To Be True 42
Messy Is As Messy Does 45
Joy Comes In The Morning 48
Do You Have A Light? 51
A Little Hope 54
Left-Footed Day 57
Looking for Mr. Right 60
Mr. Wonderful 64
A Prayer 67
Let The Sparks Fly 68
The Heart Of The Matter 72
I'm Alive 75
Not Afraid 78
Out of The Hallway 81
Duct Tape Can Fix That 84
Life IS Like A Box of Chocolates 86
A Walk On The Desperate Side 89
Rip Tide 93
But I LOVE Those Flip Flops 96
Ready To Roll-er Coaster 99
When Life Gives You Lemons 101
Are We There Yet, Lord? 103
You Are Worth More 105
A Song In The Night 108
Lost Causes 112

Title	Page
I Know Why The Caged Bird Sings	114
Don't Touch My Rose-Colored Glasses	116
I Wish I Were A Butterfly	118
Gypsy (Not Gangnam) Style	121
Ready For Spring	123
Enough	125
Cracked Pots	128
The Prodigal	131
I Have That Shirt	135
Not Enough	137
It's All In The Wanting	139
Words That I Hate For $500…Waiting	142
Finding God's Heart	144
You'll Get Through This	146
Cowboy Up	149
The Miracle Of Making It Through	152
No Good Deed	154
The Abyss	158
Where Are My Glasses?	161
Tantrum Land	164
I Have A Friend	166
The Test	168
Believe in The Magic	170
The Science of Making Toast	174
Can I Have A Cookie, Please?	177
The Only Jesus	180
A Beautiful Soul	183
Making My Mark	185
Memories Are A Treasure	189
You are Here For A Reason	191
What Can I Bring The King?	192
My Life In Boxes	194
Tunnel Vision	197
Happiness Is	199
Waiting For Summer	202
First Snow	205
Silent Night	209
Emmanuel	212
The Journey Continues	215

what to listen for and look out for, you were usually in fine shape. Life in my grown-up world has reminded me of those track-traveling days. I have skipped along the tracks of life, enjoying sights along the way, because tracks always have a way of running along great scenic routes. Rivers. Flower-covered banks. Woods. Mountains. Oceans. Covered bridges. Life can be amazing and beautiful. There are times when all is going well. Even now I can remember how it felt with the sun on my face and gentle breezes blowing through my hair on those long-gone summer days walking the rails.

I also remember falling off those tracks more than once. Ouch! Skinned, bloodied knees and hands with bits of crushed rocks stuck in the oozing cuts and scrapes. Life has knocked me down on a few occasions and left me battered and bloodied, just like that little pig-tailed girl that fell off the tracks. It hurt. It always got worse before better, because you had to clean all the rocks and dirt out, and put on Merthiolate. Remember that stuff? Apply and blow profusely. I was so happy when we discovered Bactine.

It took awhile for those wounds to heal, and some left scars. Life's wounds are the same way. Some are painful. Some are dirty. Some are hard to get cleaned of all the debris. Some take forever to heal. Thankfully, they do eventually heal, but some do indeed leave scars—reminders of not being careful, of moving too fast when you weren't ready, or not watching where you were going. They were childish mistakes then, but part of being human now.

Train tracks themselves are stern and stark fixtures. There have been times when life was as hard as the high-carbon steel alloy of those rails, and as unforgiving; times when there were no easy answers. Steel is tough, and life can be just as tough. You sometimes wonder how any good can come from some situations and circumstances. How can beauty come from ashes?

I remember how beautiful wildflowers always seemed to grow along the tracks. Even in the course ballast beneath, and under the tracks and ties, flowers would break through those crushed stones as a testament to life's perseverance. In the hardest times of my life, when all seemed crushed and broken from the weight of all I was carrying, there was a hope that refused to be buried. No matter how heavy the load, there was something that caused me to push harder, knowing I could make it through the worst and not just survive, but thrive.

And what would any self-respecting train track be without a tunnel? The length of a tunnel determines a lot about how you get through it. The tunnels in life are the same way. You may feel like you have been making your way through a long tunnel. It has been pretty dark and hard to maneuver through. When you can't see too far ahead of you, you want to take it a little slow and easy. It hurts to fall off the tracks. Some tunnels are cold places where you feel scared and alone. Other tunnels are hot and stuffy, and you feel the pressures from all around. Tunnels in life are just like that. You have no option but to go through them. You just want to make it safely out—and hopefully in one piece.

So you inch your way through, hoping that a light will appear in the distance. One good thing about walking through the darkness is that eventually your eyes adjust and you can begin to see more clearly. You don't need full clarity to begin to find your way. You just need enough sight to get you where you need to go. Ahead. As long as you keep moving, you will make it through and out. I see a light in the distance—it's the tunnel's end, and sunshine is just outside. At least I am praying that is sunshine. Dear Lord, please just don't let it be a train.

"Make me walk along the path of your commands, for that is where my happiness is found" (Psalm 119:35 NLT).

Fasten Your Seatbelts

Flying is like so many other things in life. It's hard to really understand what it's like until you do it. I arrived at the airport for my first flight, accompanied by rain, thunder, and dreariness all around. It was not the best traveling weather, but it didn't dampen my excitement. I was a little nervous. Riding in a metal canister with lightning flashing around doesn't seem like a smart move, and yet millions of us do it.

    Once on board I waited anxiously. I sat watching the rain coming down in torrents and wondering if this was such a good idea after all. Yet the next thing I knew we began to taxi down the runway. You start to breathe a little quicker and your heart races, wondering if the 400 tons of metal will actually lift off. Then it does, and it's quite a feeling. Still, that wasn't the best part. I'll never forget when we climbed up to 30,000 feet and emerged from the clouds. There was no rain. There was sunshine. Down below it was storming, but high up it was sunny. It was amazing and beautiful.

    Below the plane as far as I could see were huge, puffy, white clouds. They looked like you could actually step out and walk on them. No matter how many times I see it, it is still as amazing as that first time.

    There is a downside to flying: hitting turbulence. Most flights the turbulence is so slight you barely even notice, but then there is that unmistakable, jarring turbulence that shakes you from your head to your shoes. It can throw you around, toss you, and even make you nauseated. Welcome to my life's flight this week. I have traveled through a lot of stormy weather, had some rough takeoffs and harsh landings, but some smooth ones, too. When you're determined to arrive at your destination, you keep going no matter what the weather. I am headed some place. I am headed to where the promises that God gave me are waiting. I am aboard and ready. I have been watching the view from my window seat. We were flying along so well. I was finally above the storm and enjoying the beauty. Then it happened. I hit turbulence—*major turbulence*. This is not the kind that gives you a shifting sensation. No. It's the kind that knocks you out of your seat. Ouch! I tried to stand up, but another jolt floored me again. I told myself just stay down. So I sat there all night. Everything shook and quaked around me. I watched things being tossed around my head. I will be honest and say I was shaken. Yet then a voice comes over the speaker. It's the Pilot. He assures me that all is well. This is just a momentary disturbance. We are going to reach our destination. He who is faithful gave us that promise. I may be too shaken to stand. I may have had to crawl back to my seat. There is better weather just ahead.

When everything is shaking around you, stay on course. You may not see the sky clearing ahead or the winds calming, but rest assured that the One that speaks peace and calms the wind has it all under control. Just to be on the safe side, I think I'll fasten my seatbelt this time.

"But those who hope in the LORD will renew their strength. They will soar on wings like eagles; they will run and not grow weary, they will walk and not be faint" (Isaiah 40:31 NIV).

"He replied, 'You of little faith, why are you so afraid?' Then he got up and rebuked the winds and the waves, and it was completely calm" (Matthew 8:26 NIV).

SOME THINGS YOU NEVER FORGET

Riding a bike is one of the best things you ever learn in life, because once you know how, you never forget. I can remember what it was like learning to ride a bike. I think back to that Christmas morning when I saw that new baby blue Schwinn Fiesta. It didn't matter that it was 28 degrees, and that there were remnants of the latest snowfall still on the ground. After breakfast, despite my mama's protests, my brothers and I, along with all our friends in the neighborhood ventured out with our new bikes (or to watch those with new bikes) test the new wheels for the first time. I remember the determination I felt as I attempted my first ride. I don't remember any fear of failing or falling, both of which I did numerous times. Some falls were bloody and painful, but it didn't stop me.

I remember many other "firsts" of life. First day of school. First dates. First time driving a car. First job. First move on your own. First heartbreak. Some "firsts" take you on an emotional ride of their own. You take off with excitement, pedaling expectantly, and yet waver with trepidation until you get yourself balanced and you are able to hold yourself up and keep traveling along. Other times you waver and become unsteady, but still manage to get yourself, and your bike, upright and back on course. Then there are the times when you lose all balance. You can't catch yourself, or even put your feet down in time to prevent the inevitable crash. Those can really hurt. If you have too many of those crashes, it can make you want to stop riding altogether. Some people that have been hurt so deeply will put their "bike" in the back of the garage, and take to walking, which is safer. It can be slower, and make the trips longer, but it limits the type of injury and pain you might have to endure. I have wanted to park my "bike" out back for a while myself. I seemed to keep falling and crashing. My balance has been off, and my riding attempts have not been a pretty sight to behold. In fact they have been downright ghastly at times. They have been the kind that when you're watching someone and you know they're headed for a crash, it makes you cover your eyes and grimace. You know they're going to get hurt. You don't want to watch. It's going to be painful. Believe me, there have been some painful crashes.

Yet deep inside is the spirit of the little girl, the essence of who I was on that morning so many years ago. That same spirit so determined to keep trying, no matter what, until I conquered the pull of gravity, and mastered the art of riding, still lives inside me.

My crashes in life have been no less painful or ugly than the ones I have experienced behind bicycle handlebars, but just like in my biking, I have decided to keep getting back up and moving ahead. Sometimes wobbling. Sometimes unsteady. Sometimes skinned up, scraped, and bloody, but not giving up—never giving up. You can't stop and yet still think you will make it anywhere. Albert Einstein was among those who noticed and commented that life is like riding a bike; to stay balanced, one has to keep moving. Maybe you put your "bike" away because you got tired of the work it took to keep pedaling. Tired of crashing. Tired of getting bruised and battered. Tired of the effort your life takes to keep balanced.

It's never too late to dust off that "Schwinn" or "Huffy" and try again. The great thing about learning to ride a bike is that you never forget how. It all comes back naturally whenever you get back on one, no matter how long it's been.

So many "firsts" in life repeat themselves as "seconds." There will be other first dates, first kisses, first days on new jobs. Maybe even a first chance in a second love. So many things in life change. I am thankful that in life there are some things that will always remain constant. The skill of bike riding will always be one of them.

"We love Him, because He first loved us" (I John 4:19 KJV).

"Yet I hold this against you: You have forsaken the love you had at first. Consider how far you have fallen! Repent and do the things you did at first…" (Revelation 2:4-5a NIV).

A Little Rearview Perspective

I am no car expert, but I have learned a few things about automobiles in my life. I will spare you my knowledge of the mechanical workings of an engine, and enlighten you on something a little more basic. How about we talk about rearview mirrors? I bet you never realized some important facts about them.

There are reasons why a rearview mirror is small and a

windshield is big. A rearview mirror is important. It helps us reflect on what's behind us. We need to look back when we're driving now and then. It keeps us safe and puts things in perspective. It lets us see what's behind us. The reason the windshield is so big is because we are moving forward and need to be able to see where we're going. If you spend too much time looking back, you may miss something ahead and end up crashing.

When I look back in the rearview mirror of my life, I am amazed at where God has brought me from, and what He has brought me through. If there is one thing I have learned, it is that God is faithful. Sometimes it has been really hard. I won't lie. Life can be unfair. I have seen far too many dark times, but God has never left my side. Even in the darkest nights, I was never alone. I've made mistakes. I've trusted people that didn't deserve my trust or the faith I placed in them. I have been led down some paths where I lost my footing, but His hand reached out and steadied me. Even when I have fallen, He was the One who picked me up and brushed me off. He loves you and me that much. Even when things get tough, He doesn't leave. Even if we fall, He doesn't give up on us. Even if we take a wrong turn, or misjudge how something looks in that rearview mirror and crash into something we could have avoided, He still holds on to us.

When I look through the windshield of my life, I see that God has new things on the road ahead. I am thankful. I am thankful to know that He is a God of second chances. He gives us new hope and dreams. Everything may not be the way that we want, but if we place our trust in Him, He will bring us where we're supposed to be.

Life is truly an adventure. You never know where the road will lead you. Mark Twain said, "Twenty years from now you will be more disappointed by the things you didn't do than the things you did. Throw off the bowlines. Sail away from the safe harbor.

Catch the trade winds in your sails. Explore. Dream. Discover." Let the winds catch your sails. That is what I am doing—pushing out from the safe place to the unknown. The uncertain is where we find what we are made of.

Change takes courage. Courage is doing what you're afraid to do, and letting go of the familiar. There can be no courage if you're not afraid. We never know what we will face ahead. I don't want to live life sheltered from the dangers. I want to learn to be fearless when facing them.

God has a plan for every life. I want that plan because I know it is better than any I can make happen on my own. It takes courage to follow when you're not sure where you are headed. The hardest things and the right things are sometimes the same. Dreams are wonderful, but you have to wake up to make them actually come true.

If He gives us the dream, He will bring us to it. So today, be happy. We are one day closer than we were yesterday.

You never know what door will open today, so keep your eyes and heart open. This could be the day that door finally unlocks. I have decided to look for the open doors. I am so done kicking the closed ones.

I have even banged my head on a few, but no more. All that gets you is headaches and sore toes. I am going to trust Him to open the ones He has for me. They won't just be opening. They will fling open. When they do, I will hit that door running. To be on the safe side, you better stay clear of my doors.

"Brethren, I count not myself to have apprehended: but this one thing I do, forgetting those things which are behind, and reaching forth unto those things which are before, I press toward the mark for the prize of the high calling of God in Christ Jesus" (Philippians 3:13-14 KJV).

TRAVELING LIGHT

Maybe I should travel lighter. We headed to Youth Congress a few weeks ago, and since confession is good for the soul, I must confess something. I had ten pairs of shoes and boots with me for a two-day trip. Yes, I am serious, *and I wore every pair!* (We won't go into how many outfits made the trip!)

When I was unpacking the car after the trip, I noticed a

squirrel gathering acorns. This is probably not as easy a task as you might think, especially with tiny little hands and small cheeks. Even though they expand a little, squirrel cheeks will only hold so many nuts. He kept trying to hold as much as he could, but that only seemed to cause him to drop what he was carrying. Eventually he figured out he had too much to carry and was able to get what he needed if he didn't try to hold so much at once.

That squirrel made me think. As I was lugging my bulging suitcases back home, I began to realize that maybe I should travel a little lighter. Maybe not just on trips, but in life. While I can make perfect sense of my traveling wardrobe and the much-needed shoe selections that must accompany it, I began to realize life can be hard enough without dragging along things I don't need or trying to hold onto things I should have let go of long ago.

We all carry around baggage in this life. We carry pieces of our past. Some are light. Some are heavy. Some are overloaded. There are hurts, heartbreaks, and pain. There can be shame, guilt, regrets, failures, and disappointments. Why do I still carry them around? I thought I had tossed them long ago, but sometimes I find them still stuffed into a side compartment. They are still trying to make the trip with me, like those cute shoes I don't really need to take but just might come in handy.

No, I don't need to drag the past with me anymore. Even without taking it with me, I won't forget about it. I carry the scars of the past. I need to remember where I have been and what I have been through. Some things are only learned through pain. Yet if the lesson is learned well enough, it won't have to be repeated. For every wound there is a scar, and every scar tells a story. It tells the story that you were hurt, but you survived.

We don't need to be ashamed of the scars we wear. They are a reminder that the hurt has stopped and the pain is over.

God has mended each wound of my heart just as this body He created heals every scratch and cut. There may be scars, but in His eyes there is a deeper beauty. Every scar makes you who you are. They remind you of how strong you really are. You have endured the pain. You have the scars to prove it. Yet you have found strength in Him that you would've never found otherwise. In this life we will make mistakes. We will fail sometimes. However, I am packing as much of His grace and mercy as I can carry. I am sure I will need it. You might want to throw in a little of that, at least as much as your "carry-on" will hold. I am traveling lighter, happier. I am heaven bound, with joy for the journey. If we should find ourselves traveling together in the future, you will find me a pleasant traveling companion—but you will still have to deal with my bulging suitcases. I am always well packed. Happy travels!

Dawn

Dawn is here. It's still dark out, but across the horizon I can see the first hint of light. I never really thought a lot about what dawn actually is until the other day. Weeping lasts for the night, but joy is coming in the morning. Several times God has reminded of me this. He has then encouraged me to rejoice and look with hope "for the dawn is here." I thought that meant the morning was here, but in actuality, the morning isn't here. Seriously? I had my hopes up again and then—I have to wait *longer?* It's not morning yet? Not yet, but it is coming. The good part of seeing dawn arrive is the night is about done. I'm glad. He has been with me through the dark night, and I can never thank Him enough for that, but I'm ready for a new day to begin. He has promised great things, and day-by-day they are taking shape, a little at a time. No, we can't see how it all fits together yet, but that's what "walking by faith" is all about. Believing without seeing. Hoping when there seems to be none. Moving ahead when you're not sure where the road is going. I know that He promised to never leave us nor forsake us. That's good enough for me. He has been true to that promise. I have never had to stand alone or face the storms of life or dark nights alone. He promised that no good thing will He keep from those that walk uprightly. There are days I succeed in living right, and there are days I fail miserably. We will never be perfect, but every day we can keep trying to do better.

Judging by the way the sky looks, dawn is here. That means morning can't be far behind. Do you know what's coming in the morning? Joy! He promised that and He cannot lie or fail. So put a smile on your face and get ready. The sun is going to shine again. Let's greet the dawn with anticipation and welcome the morning with open arms. May your day be full of hope and promises coming true. The One that promised will surely do it.

"The path of the righteous is like the morning sun, shining ever brighter till the full light of day" (Proverbs 4:18 NIV).

READY TO RUN

Laces tied. Stretches done. Open the door. I'm ready to run! I love to run. Indoors. Outdoors. Uphill. Downhill. It just feels so good. Those of you that run can understand perfectly. Running takes energy, strong muscles, coordinated movements, and a drive to keep going. There's something about the steady rhythm of your feet hitting the ground, the deep breathing, and the feel of your heart pounding strongly that can bring a sense of peace into whatever chaos is going on in your life.

One thing I have learned from running is that no matter how tired I am (or think that I am) I always feel better after a good run. God bless those endorphins! Running has made me stronger and healthier in body and mind. When you get the heart pumping and the lungs working hard, the influx of oxygen has a way of breathing new life into you. My mind thinks clearer. My body feels energized.

I have learned that in the middle of a long run I sometimes get tired and breathless. My legs have felt heavy and shaky. My body wanted to stop or at least slow down and rest. Yet my heart said, "Keep going." My mind said, "Push harder." I knew I didn't have much farther to go. So I pushed myself even harder. Then it happened. I may have felt on the verge of collapse when something of a miracle took place. It's a phenomenon they call "second wind." Suddenly a new strength comes and I can keep going. It's not as hard now. I feel a rush of renewed drive as fresh adrenaline courses through my not-so-tired body. On I go.

I find God can use everything in life, big or small, to teach lessons. I know where I am in this race of life. I have run long and hard and constant. Sometimes I have run like a champion. Lately I've been tired. I was running so well over the past few miles, then it seemed that the weight and strain started taking their toll. I wasn't moving as fast or as determined or as strong as before. I stumbled. I tripped more times than I cared to count. I have even fallen down. I am dusty and dirty, but I have to get back up. I brush myself off as best I can and keep running. It would be easy to slow down and stop or turn around, but it wouldn't be right. Something whispers to me to just stop running. But I realize: how would I make it home if I do that?

I have run too many miles to quit now. I have run in the sun and rain. Through the nights, the storms, and the valleys. Over mountains. I can't see the end of this race, but I know it's near. Someone is calling. From somewhere up ahead I hear Someone cheering me on. He calls from the sideline, "Don't give up now. Don't give in. Press on." So I have to run harder. Press on past the doubts. Press on past the exhaustion. Press on past the pain. This race is not to the swift. This battle is not to the strong, but to him that endures to the end. I feel that "second wind." It's kicking in. My steps are not as heavy today. I can breathe a little easier. I am not going to be limited by what I can't see or understand because there is a new strength rising up in me. We can do this. We can finish this race and finish strong. Until we cross the finish line, we need to be ready to run.

"Wherefore seeing we also are compassed about with so great a cloud of witnesses, let us lay aside every weight, and the sin which doth so easily beset us, and let us run with patience the race that is set before us" (Hebrews 12:1 KJV).

"Do you not know that in a race all the runners run, but only one gets the prize? Run in such a way as to get the prize" (I Corinthians 9:24 NIV).

ADVENTURE TIME

Have you ever watched ducks? It doesn't really matter what the weather is or what is happening around them, they seem to adapt to whatever is going on. They swim in the sun and rain and snow. If a threat appears they can fly away to a safer place. They "go with the flow." We've heard the saying that it's better to dance in the rain than just wait until the storm passes. If you visit a duck pond on a rainy day I think you will find they have all ready learned that secret. Life doesn't stop during the rain and storm. The adventure goes on and so does life. Ducks live the adventure—whatever it happens to be.

Life is an adventure and should be lived that way. The past few years I have spent living my life in just that way. An adventure. I decided that if I have never done it, but want to, now is my chance. Adventures make us see things in different ways. Adventures exist everywhere if we take the time to look and discover what is all around us.

Nevertheless there are some days that come and you wonder how you will make it through them. No matter how hard you try, there is no avoiding them. It would be nice to be able to fly away like the ducks. Fly away from harm or things that frighten and threaten us. Fly away to a place where it's always sunny and safe. However, both good things and bad things happen to all. The rain falls on the just and the unjust. Even with an adventurous outlook in life, you aren't exempt from the pain that life brings. The hardest pain is smiling to keep the tears from falling. We wonder why we have to be hurt, disappointed, rejected, or abandoned. Yet it is the pain we go through that gives birth to a new strength.

Things in this life will go wrong, but they will work out—perhaps not as we planned, but if we put it in God's hands, it will work out as it should. "All things work together for good to them that love God, to them that are called according to his purpose" (Romans 8:28 KJV). The strongest people are the ones that have felt pain, accepted it, and learned from it. Forgive those that caused the pain, but never forget what it taught you. If you can understand this, you are a survivor. Yes, you may be scarred, but you made it through. Pain makes you stronger. Fear makes you braver. Heartbreak makes you wiser. Some of us should be super heroes.

Today is the first day of the rest of our lives. Yesterday is gone. We can't change it. We can't relive it. Who wants to? There were good times. There were bad times. We can learn from it, but it's passed. It's over. The door is closed. That adventure or ordeal ended.

Every day is a new start. I am determined to live this day with everything I have. We have no promise of tomorrow. Today let's love, laugh, encourage, hope, work, dream, believe—and repent, if need be. We may even weep, although hopefully not too much. Tears are necessary. They cleanse us within. They reduce the stress. They remind us we're human just as the mistakes we make do likewise. It would be great if every choice were the right one. But, it's not that way. We ought to learn what we need to learn and move on a little wiser.

Just as God gave ducks the ability to let water roll off their backs, He gave us the ability to let life and all it brings "roll" off our backs. Life is a gift from God, a new adventure every day. There will be good days yet to come. There will be bad days yet to come. On the good days—enjoy every moment. On the bad days—remember: this too shall pass. It is on this journey we find who we are and what we are made of. Some days we will climb high, and some days we will fall. The important thing to remember is the adventure never ends as long as we get up and keep going. Or if need be, dance.

"To everything there is a season, and a time to every purpose under the heaven: A time to be born, and a time to die; a time to plant, and a time to pluck up that which is planted; A time to kill, and a time to heal; a time to break down, and a time to build up; A time to weep, and a time to laugh; a time to mourn, and a time to dance" (Ecclesiastes 3:1-4 KJV).

Cool Running

I remember driving on cold mornings, trying to get the frost off of my windows, and there they were—the crazy runners. You could see their breath, flushed and frozen cheeks—but on they ran. You have to love what you're doing to get out of bed with temperatures below freezing (and not being paid to do it) and take off running.

It was 31 degrees this morning, sun barely up, and where was I? Yep. Running. Why? Because I can. Because it makes me happy. Because it feels good. Because running changed my life. It clears my mind and brings focus to the chaos. It has made me appreciate life. So many would give anything for a chance to walk, or run, or live just one more day.

We take so much for granted. We don't see or appreciate life. Things may be hard. I am not where I am headed yet, but I am enjoying where the journey is taking me. On the good days and the bad days. Because that's what life is made of. Everyone wants to be happy. But what is happiness? You can't travel to it or buy it, but once you find it, you can own it and wear it. Happiness is found in living life every day. When we are waiting for some great thing to come down the road, we miss out on the great things in every day. Actually today is all we're really promised any way. I don't want to take any day for granted. That's why I get excited about sunrises, because this is my day. I want to make everything I say and do count. Every word. Every breath. Everything. When the sun goes down there will be no regrets. I am giving this day all I have. I hope you will, too. So the next time you are trying to clear your frosty windshield and spot one of us "crazy" runners ahead, remember: there goes one cold but happy runner. We warm up, after the first mile anyway.

"He said, 'Come what may, I want to run'" (II Samuel 18:23a NIV).

"I shall run the way of Your commandments, For You will enlarge my heart" (Psalm 119:32 NASB).

MUD

It must have been the kid in me that thought running, crawling, and slopping through the mud sounded appealing. That's how I found myself in a Tough Mudder. I was *so* excited about doing it too. When the day arrived, it was 37 degrees. I didn't realize the race started out with a swim across an icy pond and ended the same way. It was technically too cold for going in a pond once.

We were going to do it twice. Really? I could see my breath and felt the warmth from the bonfires they had going to keep the spectators in the winter coats, hats, and gloves warm. But hey, life is about adventure, right? I knew the race would be hard. There would be mountains to scale. There was a forest to run through. There would be obstacles to climb and crawl through and hurdle. There would be water, really *cold* water. There would be mud and muck, of course. It was all part of the challenge and fun.

When I started the race I looked so cute. I had pigtails and a colorful running suit. I was coordinated even down to my headband. That stylish look didn't last long. I was soon soaked and muddy and covered with bits of straw and stickers. The first mud obstacle you encounter, you may find yourself a little hesitant. The next one you enter more freely. By the third one, you are on your belly, face down in the mud and muck and could care less. I had mud in places the sun doesn't shine. But I finished the race. Dirty and tired, but happy.

You see this life we live is a Tough Mudder. It's an adventure and challenge every day. Sometimes we can get through parts of the course upright and clean. We have a steady pace and momentum going. Then we hit a soft spot and lose our balance— falling into the mud. We get up, but we're not real pleased. It doesn't feel so good. WE don't look so good anymore. It's kind of cold, and it doesn't feel right. But we run on. Up ahead we find ourselves crawling through messy places. Not so great here either, but if you let your guard down it's easier to be comfortable.

By the time you reach the muddy pit, it doesn't even matter anymore. You go into it without a thought because you have let yourself become used to how the mud feels. Some wouldn't consider that a good thing, but you're so familiar with it now that you can accept it. You started out clean with hope and belief. You were color-coordinated and put together. Look at you now. You're a muddy mess. You are so far from how you started.

I am wondering how far is too far with God? Is there a place where He says "Enough? You just aren't worth the trouble anymore. You gladly crawl through the mud in this life without even stopping to think. There are no ends to the depths of mud you sink into." If that's what He had to tell us, we would know it is the truth. Who would blame Him? He must get tired and disappointed when we willingly dive into those low places He wants us to hurdle over. Luckily for us, He remembers we are flesh. And weak. Human.

This race of life is hard. Every day as I am running it, I try to avoid the mud and muck. But guess what? Some days I don't come out clean. Some days I am muddy. Some days there is so much mud that I can't run. I do good to walk. Who am I kidding? There are days I find myself crawling—mud covered from head to toe.

Just as the water from the shower washed away all the traces of the Mudder, His blood washes every mud covered spot in my life. His grace does what I would never be able to do. It cleans me up. And that's a good thing. The race isn't over yet. It's easier to run without having mud weighing you down.

I am stronger because I was weak. I understand my beauty because I have seen my flaws (even covered in mud). I am wiser because I have been foolish. I can appreciate laughter because I have known deep sadness. And luckily, mud and sin can be washed away.

"And such were some of you: but ye are washed, but ye are sanctified, but ye are justified in the name of the Lord Jesus, and by the Spirit of our God" (I Corinthians 6:11 KJV).

"I therefore so run, not as uncertainly; so fight I, not as one that beateth the air" (I Corinthians 9:26 KJV).

"I have fought the good fight, I have finished the race, I have kept the faith" (II Timothy 4:7 NIV).

WHAT'S THAT SMELL?

What is that smell? You know how you can smell a bad odor, and you can't figure out where it's coming from? What's the first thing you do? I would guess it's the under the arm sniff test. Admit it. I can't be the only one. If all is fine there, then you want to figure out where the smell is coming from, and get rid of it. Fast.

The sense of smell is an amazing thing. Would you like to hear about some interesting smelling facts? Did you know that everyone has their own "smellprint"? That's right. We all smell things in our own unique way. When I smell flowers, you may smell carrots. A rose may smell sweeter to me than you. It's a true fact. Another fact is that you can smell fear and happiness. Good smells make you happier. Our sense of smelling improves as the day goes on. There are only 7 main scents: musky, putrid, pungent, camphoraceous (think of mothballs), ethereal (think dry cleaning stuff), floral, and minty. We each have scent "blind spots" or certain odors that we can't pick up on.

You actually smell with your brain. Scents bring back memories—good and bad. I watched a documentary one time about the world's smelliest animals. It was quite enlightening. I was glad we didn't have "smellevision" for that one. One interesting fact they shared was that when you smell something, particles of the odor actually get in your nose. That is really gross if you think about it too long. We have 350 odor receptors. It seems like a lot, but mice have 1300. We breathe in innumerable molecules all day long with our odor receptors, but it's only when a certain scent makes us happy, pleases, or irritates us that we notice it. Your nose can smell directionally, helping you find where an odor is coming from.

So I wonder where that smell that is bothering me is coming from?

We were out riding our bikes one evening and saw a baby skunk toddling around some beehives near the road, probably looking for some drops of honey. It was so stinking cute (no pun intended) that we had to stop. The next thing we knew there were two others that came sauntering up. Now a sensible person would probably think twice about cohorting with animals of questionable behavior that spray potent and long-lasting odor in dangerous situations, but sensibility can be overrated. You can miss a lot of adventures in life from being too sensible. Once we figured the mom was nowhere around, we relaxed and spent the next hour taking pictures and following the babies with an unhindered fascination. "How young can they spray?" I asked when a moment of clarity passed quickly through my head and left. With no way to Google out in the country, we just ventured ahead, throwing caution, but hopefully not our good "scents" (pun intended) to the wind. At first the little ones wandered around unafraid, even coming so close that they walked right past our feet and right up to the camera lens. They seemed to have a little trouble seeing, and they just didn't understand there were things to fear. Yet.

As time would have it, the natural instinct of being a wild animal kicked in and we found out that, yes indeed, baby skunks can spray. We were luckily a few feet away when we discovered this fact, so we were able to avoid multiple tomato juice baths. Spraying their scent could have just been a reflex that they didn't even realize they had or were doing. They may not have even realized that smell came from them. The odor was probably familiar. I am sure Mama had a smell to her. They might even find it comforting. They were, after all, skunks.

This encounter made me think. I realized something. I realized the thing that actually was stinking was me. Ouch! Or should I say "P.U."? I have spent days sauntering around just like those little skunks that appeared so innocent and appealing. But appearances can be deceiving. We all put on a face to present to the world—real or pretend. There is an old saying that says if it looks like a skunk and walks like a skunk and smells like a skunk, it is probably a skunk.

Those skunks weren't trying to hide what they were. We may not intentionally hide what we are either. But we may try to hide the struggles, the battles, the shortcomings, and the failings. The things that make us feel like we aren't what we should be or that we fall short—or that we stink. The truth is that no matter how good we think we are, how much we try to do the right thing, we will always fall short. We're human. We will never be perfect. "All of us have become like one who is unclean, and all our righteous acts are like filthy rags..." (Isaiah 64:6 NIV). No matter how hard we try and no matter how much we do, it will never be enough on our own. We will have days that, just like those little skunks, we will stink—on purpose or accidentally. Sometimes we can hide it from others. Sometimes we can even hide it from ourselves. However, bad odors have a way of getting worse if ignored.

Thankfully there is a remedy. Like the tomato juice helps remove the "skunky" stench, we can be washed clean from the things that make us stink. He has us covered, by His blood. "Unto him that loved us, and washed us from our sins in his own blood" (Revelations 1:5b). We may make mistakes. We may sometimes miss the mark, but all is not lost.

I am so glad for His mercies that are new *every* morning. It doesn't matter if you fell down yesterday, what matters is that you get up today and keep going. Every day is a fresh start. Another chance to do better. Start every morning with a clean heart and end each day knowing that you did your best. On those days when you don't quite get it right, just remember that His mercy never ends, and His grace will never leave us. And now for some reason, I feel the need for a shower.

I YAM WHAT I YAM

"I yam what I yam and that's all that I yam!" Popeye said that, but I am laying claim to it today. I have made mistakes—plenty of them. I have owned them every one. I am undoubtedly human. But I have never set out to intentionally deceive anyone. I can't think of one time when I have taken advantage of or used anyone either. I am *not* perfect. God knows that. I am just a sinner saved by grace, trying every day to make it through. Some days I do better than other days. I always try to keep a positive attitude, see the good in people, and believe the best in people.

Have you ever started looking around you? I did that recently. Mistake number one. Call it momentary weakness. We all have those lapses when we take our eyes off where we should be looking. But sometimes you have those moments where things wear down your defenses and better judgment. You get so tired of all the fake people. I am sure you know some of them. They talk the talk, and sometimes even look like they walk the walk. But appearances can be deceiving. The old wolf in sheep's clothing. I have met a few of those. I've even met a few goats masquerading as sheep. I knew a farmer that had an orphaned baby goat he had raised with his lambs. While "Spike" was close enough in species to get along reasonably well in sheep land, no matter what he did or didn't do, he would never really be a sheep. He may have lived the same life they did, but when he grew up, he was a goat in looks and actions.

I know we are all human. But being human shouldn't be a cop out for doing wrong. Do you know what I want? I want people to be what they say they are. I want people to be real. Real. Is that so much to ask from any of us? Just be what you profess to be. I want people to look me in the eye and speak the truth. I want to be able to trust and believe in people.

I know they are out there. I know some. There may not be a lot of them, but those are the people I want for my friends. Those with integrity. Those that won't cause you pain so that they can gain something. Those that won't compromise their beliefs for any reason. Those that will help me be a better person. Those that will pick me up instead of step on me on their way to where ever they're going. Those that will be there through thick and thin, good times and bad times. That's the person I want to be. The friend I want to be. When it's all said and done, I want to be real. I won't ever be perfect this side of heaven, but I will be real. In my good and my bad, take me as I yam—but let me be real.

"But by the grace of God I am what I am: and his grace which was bestowed upon me was not in vain; but I laboured more abundantly than they all: yet not I, but the grace of God which was with me" (I Corinthians 15:10 KJV).

A Reason To Be

Why do we think what we say and do doesn't matter? It does. Your actions affect someone else. Sometimes you can't see how it does, but trust me, it does. It's like when you toss a little stone into water and see the ripples, everything you say and do causes a "stir" in the "waters" around you. Water in a lazy stream courses gently along for miles with little noticeable changes. But water is powerful and can change the land around it. The Grand Canyon is a prime example of that theory. Water may look like it's not doing anything more than flowing along, but let it reach a different stream bed full of rocks and a change in the current. It will turn from a gentle stream to the raging rapids.

We can do that in someone's life. All it may take is a simple statement or action to change someone's world. That can be for the good or bad. What you do matters. What you say matters. No one is perfect. No one is faultless. I admit it. I'm the man in the mirror on this one. I have had to take a long, hard look at myself more than once.

But why should we care? Am I my brother's keeper? Actually, yes. Is it your fault someone falls or fails? No, but just like one little straw broke the camel's back; your contribution could make the difference in whether they stand—or fall. We try to keep our eyes off of others, but let's be honest. None of us do—not completely. In the big picture, we need each other.

I want to be who I present myself to be. I want to be better every day. I don't always succeed. I am me and you are you. The good and the bad. The success and the failure. The wins and the losses. The butterfly trying to get out of the stinking cocoon. If I have fallen short of what I should be, forgive me. I am trying. This week I was reminded again how crushing it can be when we find out someone we look up to is—God forbid—human, too. That being said, we need to realize that who you say you are, and what you say you are, is who you should be, or at least should try to be. Be someone's reason to keep fighting, not the reason they quit.

"Let us not therefore judge one another anymore: but judge this rather, that no man put a stumbling block or an occasion to fall in his brother's way" (Romans 14:13 KJV).

Too Good To Be True

If you've ever been awake in the middle of the night, roaming the hundreds of television channels for something to watch, I am sure you have come upon those nifty products. You know the ones. The gadgets that are so unbelievably fantastic and life-changing that you can't live without them. You used to only be able to get them by ordering, but now you can find them right in the stores. Imagine. I must admit that I have purchased a few of the more sophisticated products. I have to tell you that the Pancake Puff Maker exceeded my expectations. I love that thing! It was worth every penny. But that was probably the exception to the rule. I have other purchases that didn't quite live up to all their hype. They were just too god to be true.

In life, when something looks too good to be true, it probably is. I found that out for myself on many occasions. You know how you just can't tell some people something and they'll accept it? No, they have to test it out and find out for themselves. They need to push the boundaries, be their own boss.

I blame the fact that I've spent so many years hanging out with two and three year olds all day that some of their dispositions have rubbed off on me. Yeah, that sounds about right. All day long I see them try to figure out this big, confusing world we live in. As if it wasn't enough to figure out how to turn on the faucets, open doors, put on their shoes, they have to learn to get along with all these other little people (and big people) they share their days with; people that have different ideas and different desires and different plans than them. Their own personal agendas come face to face with everyone else's personal agendas. That's the story of everyday life in every one's world. My needs, my wants, my desires, my beliefs meet your needs, your wants, your desires, your beliefs. It makes for endless complications.

We are all trying to make a life for our self. We want a good one—or a great one. We are all out on our own journey to Destination Happiness or Success or Fame or Survival. We are all headed somewhere.

Along the way we will come upon things, and even people, that seem too good to be true. So many of those As Seen On TV products look so good. They look like they could greatly improve your life, make things easier, or just do really cool stuff. You will meet people that are just like that too. They are smart, and witty, and fun. They seem to have it all together. The packaging is so impressive. You think they can really enrich your life.

At first it's great. You look forward to every exchange, every meeting, and every encounter. But just like that old Ginsu knife tucked away in your kitchen drawer, there comes a time when you realize they're not as sharp as you thought they were, not as skilled as the commercial presented, and probably not able to actually do what you thought they could. They are not what they seem. They're too good to be true. That is a sad day when you realize that. I hate when that happens. Sometimes you feel like you were scammed. You feel kind of ticked off. You toss it into the back of the drawer or in the trash.

Sometimes the disappointment is overwhelming. You put a lot of investment into having it in your life, and now you just feel cheated. You feel like you were taken advantage of. There is a good chance you were. That makes you less trusting and more cautious, and sometimes bitter. In a world that embraces the cynical, it is easy to lose hope when you are disappointed by life. We all have been at one point in our life or another. It can leave you feeling lost and empty. It's easy to doubt and wonder who and what you can put your faith in.

But don't lose hope. Even though my belief and trust has been crushed many times like ice in a Magic Bullet, I know there is still a reason to keep believing. "For I know the plans I have for you, declares the LORD, plans to prosper you and not harm you, plans to give you hope and a future" (Jeremiah 29:11 NIV). He has plans for me. He has plans for you. Good plans. Others may use and abuse us, but it doesn't change His plans. "You intended to harm me, but God intended it for good to accomplish what is now being done, the saving of many lives" (Genesis 50:20 NIV). Joseph's life was anything but peaceful. But he knew the one thing that we should remember. No matter what we face, and no matter what is done to us, even behind the worst that life can bring, God is working it all out for our good.

Things may not turn out like you thought they would. Things may not work out like you expected. But, if you have placed your life in His hands, they will turn out as they should. God cannot lie, and He promised "But as it is written: Eye hath not seen, nor ear heard, neither have entered into the heart of man all that He has prepared for them that love Him" (I Corinthians 2:9 KJV). He has things for us much better than we can even imagine. "No good thing will he withhold from them that walk uprightly" (Psalm 84:11b KJV).

The world has a lot to offer that seems too good to be true. In most cases it probably is. Put your trust in the One that not only gives good to those that love Him, but gives His best. No need to have a money back guarantee. He is as good as His word.

"For the word of the LORD is right and true; he is faithful in all he does" (Psalm 33:4 NIV).

Messy Is As Messy Does

What a mess! Would you look at this dirt? I don't know how I'll ever get the stains out. There are wrinkles and smudges. There are tears and rips that look pretty bad. It's ready for the rag bag. No one should ever go out in public like this! This looks beyond any hope of repair. How could I even think about venturing out looking like this?

If you looked in the same mirror I am looking in, you would probably be confused. The outfit I'm wearing is not in any way messed up. In fact, it is put together quite nicely. As far as anyone can see all appears right.

But to be honest, I am a hot mess. That would be a good way to describe it. Not familiar with that term? They define a hot mess as being when everything looks good on the outside like you've got your life together, but really things are far from together. That is what is so deceptive. We can dress ourselves up. We can fix our hair just right. Say the right thing. Do the right thing. To anyone looking, we seem fine. We "got it going on"— on the outside. We can look picture perfect, but, if they could look on the inside, they would see a whole different picture.

Reputation is what others perceive you to be, but character is what you actually are. It is the thing deep inside that drives you toward the decisions and choices you make. It is the core of who we really are. Character is what we are when no one else is looking, except when we face the mirror. Yet that person already knows it all; even what can't be seen in the reflection staring back at us. I know what others can't see. I know the thoughts, the mistakes, and the shortcomings that are a part of my past. I know the temptations and weaknesses I struggle with and battle day by day, hour by hour, and sometimes minute by minute. I know. You may not see any of it, but when I look at myself, I know. I can see it. Some days it's not a pretty sight. It can be downright ugly.

Do you know what I'm talking about? Even when we try our best, our best never seems to measure up. We all fall short. Whether we admit it or not, we can't pass a mirror without taking a look, even a quick one. Call it human nature or just a habit; we are a vain and curious species. For so long I hated looking in the mirror, not because of who I was, but because of who I wasn't. I wanted to be stronger. I wanted to be better. I wanted to be who I was made to be. Those aren't the things you see in a mirror, and yet when we face ourselves there, those are the things that call out to us. The flaws. The imperfections. The defects. Others may not see them, but we do. They can hurt and even cripple us. They can keep us from reaching who we were destined to be. Yes, there has always been a plan for your life. We try to do it our way too many times. It doesn't always turn out the way it should. That's where the mess can come into the picture. But when we turn it over to Him—not just our dreams and plans, but our shortcomings and mistakes and failures—He can work miracles. "We can make our plans, but the LORD determines his steps" (Proverbs 16:9 NIV).

Instead of seeing what's wrong in your life, try to look for what's right. So you're not who you should be. But you aren't who you used to be. You may have a long way to go, but every step gets you closer. You haven't gone so far that you can't get back. You may have messed up so much that you feel you are starting all over. That's ok. It just takes a first step. Or first look.

Sometimes it is a hard look. But once we face the truth in the mirror, and in ourselves, it gets easier. I don't want to avoid the mirror. I just want to be able to look at myself honestly. I don't want to cringe for bad choices in life...or fashion. After another look, I don't think the damage is unfixable. The mirror should be our friend, not our enemy. I pray your reflection is kind, and if not, then I hope it's something a different perspective can make better.

"For now we see through a glass, darkly; but then face to face: now I know in part; but then shall I know even as also I am known" (I Corinthians 13:12 KJV).

Joy Comes In The Morning

I don't know how many sunrises I have watched these past few years. There have been too many to count. Every one unique in its own way, and I have hundreds of pictures to prove it. I have been blessed so many mornings as I watched those first lights of morning breaking across the eastern skies. I'm an early riser, so I'm usually up before the sun and enjoy my morning tea as I wait for it to make its appearance.

But I have seen quite a few sunrises that have come following some long, dark nights. I have lived for God most of my days, but have not been immune to having to face the night. You would think it could be avoided but it can't. It's not a punishment either. It's just a part of life just as the night is a part of day. It can be scary at night if you're afraid of the dark or what may be in the dark. Everything seems larger and scarier in the dark. Sometimes the darkness is almost tangible. It can be lonely at night. So quiet. So solitary. So dark. It makes you feel like you are the only one alive. Alive. Alone. Awake.

Sometimes sleep just doesn't come for whatever reason. Pain. Fear. Sickness. Worry. Heartbreak. We all have to face those nights. But I can tell you from personal experience, there is a lot to be found in the darkest nights. I have found that not once has God ever left me alone. He that keeps me never sleeps or slumbers. He won't nod off when I need Him most. No, He is there for the duration. He has heard every whispered prayer—every desperate plea. He has caught every tear I cried. He knows every longing and dream. He knows it all because He has been there with me. Through it all. Just Him and I. And it's been enough.

Some would think there were times I had no reason to sing, but I lifted up my voice to Him and He has given me a song in the night. He has covered me in wonderful peace like a warm blanket. His perfect love really does cast out all fear. It can't be much longer now.

Time goes slowly when you're waiting. Just watching the darkened sky. Then you see it. There is that first line of light across the horizon. Your heart beats a little faster. You breathe a little easier. You have made it through the night. The dawn is coming. The dawn brings a new day—full of new mercies, new promises, and new dreams. I remember He promised

"...Weeping may last through the night, but joy comes with the morning" (Psalm 30:5 NLT). And He always keeps His promises. Always.

Do You Have A Light?

Everything looks different in the light of day. Have you ever woke up at 4:00 AM and thought about something you were facing, something you were going through, or something you just weren't sure about? Things always seem worse at 4:00 AM. Bigger. Darker. Scarier. Some days you have it all figured out. You can take care of it all. You've got this. No worries. Then night comes. Things don't look the same in the darkness that's all around you. They take on a different shape in the obscurity of the midnight hour. You don't feel so secure now. Doubts surround you in the pitch-blackness. You knew where you were headed a few short hours ago, but now? You can't even find your way across the room, let alone out of your circumstance. How could everything seem so possible and sure just a little while ago?

Now things look so uncertain. What is it about waking up and realizing you are in the dark that can make you feel so confused and startled?

Remember when you were little and had a bad dream only to wake in the darkness disoriented and frightened? You look around the room, but it all looks so different. What is that over by the door? Is something moving in the closet? You want to call out, but for some reason no sound will come. So you lay in the dark, clutching the blanket to your face, not covering your eyes, because as scary as the thought of what might be lingering in the shadows, it's even scarier to think of what you might not see coming.

And that's how it is at 4:00 AM. You can't see clearly what's going on. It's all distorted. But instead of looking away, you keep trying to stare into the darkness—trying to make sense of it all. When you can't see clearly, you can't make accurate judgments about anything. Nothing is recognizable when you try to make sense of it in the darkness. Your eyes just can't focus without the light to really see things as they are or will be. The shadows make it appear that something really scary is waiting nearby. There's a big thing looming across the room—or in your life. It may look like monsters, or failures, are coming at you. But switch on the light and things suddenly become clearer. That wasn't Frankenstein lurking in the corner in the dark. A flip of the light switch and it turns out to be SpongeBob.

If you are facing a 4:00 AM fright fest, don't let the fear you feel tear your hope out of your hands. Clutch your blanket a little tighter and close your eyes if you need to. Call out to the One that can push the darkness away. Just stay keep calm for a little longer, and the Light will shine. And in His light, you will see clearly, and all doubt and fear will disappear like the shadows slipping away in the first rays of the morning sun. Everything looks different in the light of day, and that is a good reason to get back up.

"For God hath not given us the spirit of fear; but of power, and of love, and of a sound mind" (II Timothy 1:7 KJV).

"The light shines in the darkness, and the darkness has not overcome it" (John 1:5 NIV).

"If I say, 'Surely the darkness will overwhelm me, and the light around me will be night,' even the darkness is not dark to You, and the night is as bright as the day. Darkness and light are alike to You" (Psalm 139:11-12 NASB).

A Little Hope

Do you know what the worst thing in life to lose is? Hope. No matter what else has happened—if you can hold on to hope— you can believe that there is the possibility that things will still work out. Some days it's easy to hope. You just know everything will be all right. But what a difference a day can make.

Look at those that had followed Jesus. For 3½ years they had seen miracle after miracle. They had heard His words. They had seen Him bring peace and healing and life. Hope. If anyone should have had hope, it should have been them. But in one day it all changed. And where were they now? Some were gathered together mourning. Most were hiding. One was dead. They had walked with the very God of the universe. One day changed all that. They felt hopeless.

Why? He had told them what was going to happen. He also told them He would rise again, but sometimes the things we see cause us to doubt what we know in our heart. They had seen the One they trusted in die—and with Him, their hope died. Yet if they could have held on to a little bit of hope—just believed a little—the dark night wouldn't have been so dark. Things would have still been sad, but not hopeless. Hope wouldn't have changed their situation, but it would have changed them.

That is how we should try to live. We should always hope. It didn't matter whether they lost hope or not, because He still did what He promised. Things may not be exactly what you are hoping for yet. It doesn't matter if some days you can't see where He is leading. It doesn't matter if you can't understand how things will work out. Some days may not look as hopeful as other days. He will still do what He's promised. Keep believing and hoping.

Those dark nights passed and they found the tomb empty. Their Hope was alive. It may still be night in your life, but morning is coming. God will roll the stone away and His resurrecting power will bring all those things He has planned to life. It is like the tulip that waits to break through the thawing ground and feel the spring sun and live. We wait and hope for the sun to reach us, and for a chance to live. Hope lives, even when it is cold and dark and you are waiting. I pray that as each morning dawns, you are filled with a hope that never dies.

All it takes is a little hope....

"May the God of hope fill you with all joy and peace as you trust in him, so that you may overflow with hope by the power of the Holy Spirit" (Romans 15:13 NIV).

Left-Footed Day

I've had a lot of left-footed days recently. Do you know what that is? It's like walking around with two left feet. It doesn't sound like a big problem—until you take a step. You are thrown off balance. Every step feels like a stumble. And then you find yourself flat on your face. So you get back up and try again. Now that you know what the problem is, it would seem you can compensate for it. That's easier said than done. The world just feels a little different when your balance is off. The task of getting anywhere now becomes a problem. This is how life gets sometimes. The way we are used to walking is not going to get us where we need to go. We need to learn how to walk a "new" way if we are ever going to reach our destination.

Every step takes concentrated effort. These are the times in our life when God is showing us that despite our best efforts, there are some things that can't be changed. But we can change how we look at the situation, and how we respond to it.

So you may not be walking as fast as you would like. Maybe there's something God doesn't want you to miss. You're unsteady on your feet. Maybe God wants you to reach out for His hand to help hold you up. You don't like feeling that you can't do it on your own. Maybe God is saying "Trust Me. I know what is best for you. I've got your back. Just hold My hand and I'll help you."

It has been said that a bumblebee defies the laws of physics by flying. Scientists have argued that a bee should not be able to fly. They have taken their formulas and calculations and from all they can determine, it simply should not be able to fly. But yet it does. All the time. We are like that bumblebee. It looks like there is no reason for us to try. Any attempt we make will be clumsy and unsuccessful. But like the scientist, our assumption would be all wrong. We can fly. Or at least keep walking.

So today, make another attempt at it. You may not arrive with a lot of charm and poise, but—with His grace—you are going to make it. Hope you make it through your day, gracefully.

"The godly may trip seven times, but they will get up again. But one disaster is enough to overthrow the wicked" (Proverbs 24:16 NLT).

"A man's heart deviseth his way: but the LORD directeth his steps" (Proverbs 16:9 KJV).

Looking for Mr. Right

I have been on the lookout for a new horse. I found a site with eight steps to buying a horse. Who knew there was so much to consider when buying a horse? I will save you the long list and summarize the main points. You have to consider whether you can afford one. They are not cheap to buy or care for. You need a place to keep them where they will be happy and healthy. You have to decide what breed appeals to you. There are so many different breeds that have different abilities and dispositions and purposes. The big thing is the commitment. They can live a long time and you have to decide if you are willing to commit to the relationship for the long haul.

That brings me to a new train of thought. Way back in the Garden of Eden God said that it wasn't good for man to be alone. So that explains why the world is on the lookout for the perfect match, their better half, their soul mate. Looking for a mate is a lot like looking for a horse. There is a lot to consider. If you don't know a lot about horses, you may have trouble finding your way in "horse land." You may not understand their behaviors or dispositions right away. The single world is a strange land, too, where it is hard to interpret and learn the language, and even harder to understand the customs. You may be looking for someone that speaks the same language you do. Or at least appears to be from the same planet. Let me tell you—it can be a "Where's Waldo" experience. I speak the truth from personal experience. You feel like you're searching everywhere and he's nowhere in sight. Maybe you're on the wrong page. Or in the wrong pasture. Am I really equating finding Mr. Right with choosing the right horse? It certainly looks that way.

The knowledge used for horse buying can be helpful in other areas in life—specifically finding that "special" someone. You have to think about what the cost of having one will be in your life, monetarily and emotionally. It will cost you. Most good things in life cost something. At times the expenses will be high. If it's what you want, then the cost will be worth it.

You have to consider the horse itself. You will be committing to its care for many years. There has to be a bond if it is going to be a happy relationship. You may feel a connection. That is a foundation to build on. You have to build a trust between horse and rider. It will take time, probably years, to learn all there is to know and deepen the relationship. Isn't that how love works? You start out with faith that this will be the love you have been waiting for. You can't know how it will all turn out, but you have faith that this person will be all that you believe they are. The trust is built little by little.

No one wants to be alone I don't believe God intended us to be alone. There are periods in our life when we will be, but I don't believe He wanted us to stay alone always. It would be lonely.

Although it seems impossible, you can with someone and are the loneliest of all. No one wants that. Been there. Done that. Not again. So I set out looking for a prince. Maybe you are, too. A happily-ever after. A fairy tale-that's not a tale. A real love story. I was like Cinderella waiting for the prince to bring my glass slipper. The only problem was—those showing up were bringing the wrong shoe. Go figure. Some look like something you could love, but they just don't have the right "feel" to them. Some just will not fit. No matter how hard you could try, they would just make your life too uncomfortable. Others have potential, but they might end up leaving you hurting from not being the right "fit" for you.

One thing I did while I was waiting was pray and make a list. I told God the things I wanted in the man He had for me. I put it in my Bible. I waited and trusted and believed.

God has this. We can trust Him. He will write a love story worth reading. And living. I want His best. You want His best. Everyone wants someone that will sweep you off your feet and make you the princess of his kingdom. Or at least the queen of his heart. Have you gazed up at the night sky and wished for love? I can't be the only dreamer around. I wished on a star he would find me. I wished on a lot of stars. I wished that he would hurry. We know that when we trust God His timing and plan is perfect. We may not understand the wait, but one day it will make sense. My foot got a little cold and I sure got little tired of hobbling around with only one shoe on. But this Cinderella hobbled on. It was worth the wait. God had the "perfect fit" for me. If you are waiting for your Prince Charming to come, keep praying and preparing. Maybe he'll come riding on the perfect horse. You never know.

"Whoso findeth a wife findeth a good thing, and obtaineth favour of the LORD" (Proverbs 18:22 KJV).

MR. WONDERFUL

I'm looking for Mr. Wonderful. Have you seen him? I know he's around here somewhere. I've seen him in my dreams. Wished upon a star for him. I prayed that God would send him. Maybe if I tell you about him, you'll know where he is. He's that hunky guy that takes my breath away when he laughs. His smile makes me feel warm all over, even from across the room. He'll bring me flowers for no reason. He'll send me texts just so I know he's thinking of me. He'll say stuff that only he and I get, just to make me smile. He'll say secret things that make me blush. He won't care if I play my favorite songs over and over and over.

He'll tell me how pretty I look and I'll believe him. He'll be proud to introduce me as his girl. He'll keep me safe. I won't have to be afraid as long as he's watching out me. He'll be strong and manly, brave and true, a real man—a one-woman man. My heart won't have to be afraid that he'll break it, because he'll treasure it as his own. He's not perfect, but that's OK because neither am I. Yet he's perfect for me. He'll appreciate all I have to offer. He'll realize just how lucky he is to have me. We'll talk about everything, and nothing. He'll get my jokes—not everyone does. He'll make me laugh—not everyone can. He'll give me a reason to smile by just entering the room.

Sometimes we'll laugh and pray together. Sometimes we'll cry and pray together. There will be times we disagree. There will be days we wonder what are we doing? But come what may, together we will be. He won't just say he loves God—he really will with all his heart. He'll love going to church (And even holding my hand there.). He'll teach me things about God and the Bible that I never knew even though I have lived for Him all my life. He'll pray with me and stay with me in the good times and the bad times, because it'll be a marriage sent from heaven. We'll walk on the beach in the moonlight enjoying beauty that only God could create. We'll camp far from the busy life under a blanket of stars and be so thankful that in a crazy, crowded world, God brought us together. He won't ever have to worry whether someone loves him or cares about him, because I'll spend every day of my life making sure he never doubts it. I'll wish that I had found him sooner so that I could've loved him longer, but I'll take every day I get with him.

Call me a hopeless romantic. Call me a dreamer. I still

believe in the fairy tale. I still want the happily ever after. And with God's help and blessing, I'm going to get it. After all, He's writing this love story. So if you see my Mr. Wonderful today, send him my way—and tell him I'm waiting.

(This was written before I met my Mr. Wonderful.)

"Delight thyself also in the LORD; and he shall give thee the desires of thine heart. Commit thy way unto the LORD; trust also in him; and he shall bring it to pass" (Psalm 37:4-5 KJV).

A Prayer

I am praying for you. I am praying that you are a real man of God. The real deal. That you are what I see. I pray that every day you are drawing closer to God. I pray you are growing stronger as each day passes. I pray God's peace always surrounds you, even when the storms are raging. When your work is hard and chaotic and stressful, I pray He will hold you up in all that you have to do.

I pray He blesses all that your hand touches. I pray you will have wisdom in all situations. I pray you will find His grace is always sufficient and His mercy is never lacking. I pray His word speaks to you every time you read it and comes to your mind whenever you need it. I pray that you will be compassionate and kind; someone that hasn't let life's disappointments make him bitter, but one who can understand what it's like to be down and offers his hand to help someone else up.

I pray that you will love God more than anything, because then I know my heart will be safe with you. I am praying for you every day. I'm waiting for the day we finally meet. Until then I'll keep praying for you. When God says it's time I know you will be the answer to my prayer. You will be chosen especially by God for me and sent directly from heaven to my heart. So until then I'll keep praying for you. I am not enjoying the wait, but I believe with all my heart, you will be well worth it. "I thank my God, making mention of you always in my prayers" (Philemon 1:4 KJV).

LET THE SPARKS FLY

Love is electric. It's like lightning striking the heart. I have had those moments. Yes, more than one. Admit it: at one time or another in your life, it happened to you. That spark. Old sparky. God didn't design us to be alone. He didn't. Having someone beside us was His plan from the beginning. That was the original plan in the garden. He made the man. He made him a woman. Until we find that special someone, we have an innate sense in all of us that keeps us in that "looking" mode. Once married that changes—or it should, but that's another thought for another day.

There is a place in us only God can fill, but there is another place that longs for someone else. The searching isn't always even a conscious mode. I learned a lot recently in my Psych class about states and altered states of consciousness. "Consciousness" is the awareness of the sensations, thoughts, and feelings being experienced at a given moment. That sums up love—and lust.

I am no electrician, but I can see a parallel here between love, lust, and electricity. Think about it. Can you remember when you met that "special someone"? There was a jolt of awareness that this could be the *one* you've been waiting for. Or maybe it was just a "current" trying to conduct its way through the wires to where it can make a connection. Maybe you'd never even spoke to each other yet. Maybe you didn't even know their name. But when you looked across the room—you felt it. Or you started talking with someone and felt like you had known them forever.

We have this wacky wiring I call the "love circuit." Think of your heart and other hearts looking to make that "love connection. Feelings or attraction is the "electricity" that flows. You can't see it. But it's there. Volts are the measure of the electricity's power...the force. Sometimes there is a strong current forcing its way from your heart. Other times it seems like you are a closed circuit. Not much seems to be happening.

When everything is at rest, the force is waiting to be unleashed. Voltage is always around, even if no electricity is being used. You may not even realize, but those desires are never far from the surface. They can flow through the wires at any given moment, and all they need is a conductor. If there is a mutual attraction, the "flow of feelings" is conducted with little resistance. That's what makes electricity so dangerous. You can touch something that appears safe and harmless, and find yourself knocked off your feet. You have to be careful because the human body is a good conductor for the force to be unleashed through.

A heart needs a good insulator, or you can be hit by an "electrical" surge when there is a sudden increase in "love currents." Sometimes you are hit with a bolt of lightning of feelings and desires. If you're not prepared, it can allow damaging "volts" to rush into the unprotected heart. It can cause you to overheat, burn out, and cause major damage. Power surges and heartbreak do that. When you get caught in a "power surge" there is a chance that destructive forces may knock you off your feet. You had better know where and what you are grounded on.

You have to be careful that what you are trying to connect to is well-grounded, too. That will reduce the risk of being shocked from having your feelings displaced by faulty wiring. You don't want to connect with someone that doesn't have their feet firmly on safe ground. You want someone that will not only be with you through the tingly sparks of new love, but someone that can stand with you when life's stormy times make you feel like lightning is striking all around you.

When you try to hook up wires, you better know what you're doing. When you try to hook up hearts, you better know who you're dealing with. People, just like wires, can be misleading. You don't always know what you are handling. It may look harmless and docile, but it can be hiding a force to be reckoned with. You have to be careful. Make sure you know what you are dealing with. Lust is usually high voltage, so be careful when you are handling it. It's better not to.

When you start dabbling with dangerous levels of electricity, it can be deadly. So can lust.

How do you know that your wires are safe and up to code, that you aren't going to short circuit or blow the whole fuse box out with what you're doing? Start with Electricity 101. Or Love 101. No matter how badly you want or need to find that someone, you can't cut corners, and you can't let your guard down. You have to pay attention to what you are seeing. You should pay attention to what you are doing and exactly what you are feeling. When you get down to the bare wires of attraction, you don't want shocked, or in the worst-case scenario—electrocuted. You don't want to have to pick yourself up off the floor from touching something you shouldn't have or come to after being blindsided by the force of something you didn't know how to handle on your own. Proverbs 4:23 says to "Guard your heart above all else, for it determines the course of your life." Your heart is valuable. It contains your hopes and dreams and passion. It is how you connect to God and everyone else. Make sure you keep it safe from unnecessary shocks. We all want a love that leaves us breathless and even "shocked" at the intensity of the feelings and attraction. But, it also needs to be grounded and insulated from the things in this life that can fray the very wires keeping you connected. Remember that one day you will find that one you have been looking for. Chances are—when your eyes meet—there'll be a spark.

The Heart Of The Matter

Smile. Just keep smiling. No one will know that deep inside you are dying. You are barely holding on. It is taking everything within you to hold back the tears. You will fall apart if you don't keep your guard up. Your heart is broken. It's nothing but bits and pieces. It's been trampled and crushed so many times that you're pretty sure it's not only unraveling, but in shreds. There can't be much left to keep it together. There is nothing to even hold all the fragments in place. God knows you are trying to keep it all together. You are trying to make it through another day without completely dissolving into a weeping pile on the floor. So you smile. You have to.

If we could see into someone else's heart, what would we see? You might be surprised. I'm not talking about the heart that is pumping and beating every second. We all have one of those. Those are easy to view if they need to be seen. But we all have another heart. I'm talking about that "heart" deep inside of us that "feels." Love. Passion. Joy. Dreams. That is the place where dreams are born and shattered.

We live in a world where everyone is looking for love and acceptance and hope. If we could look inside everyone we meet, would it change how we view the world? Oh, yes, I am sure it would. We put on a brave face sometimes. We smile when we are hurting. We laugh when inside we are breaking. We pretend we have it all together, so no one will really know that it's all coming apart. If we could see in each other's hearts and see the pain and fear and hurts that we carry, would it change how we see each other?

We all fall down. We all make mistakes. We all are struggling at times. We are all humans just doing our best—and sometimes not doing our best. But if we could see each other that way, couldn't it change how we treated each other? I hope it would.

Have you heard the term "black heart"? I have known a few. I am sure you have, too. They're that person that cares about no one but themselves. Once you have encountered one of them, you never leave the same. Some are easy to spot. Others are not as easy to recognize. They appear as friends. But eventually their true color shows through. Black. But by that time your heart is usually black and blue from the beating it took at their hands.

On the other extreme is that person that "wears their heart on their sleeve." I have been known to do that on more than one occasion. You have no trouble knowing exactly what I am feeling or where you stand with me. I may not say it, but I sure can't hide it. They say that if you look a person in the eye, you can tell how they really feel about you in five seconds.

The Bible tells us to guard your heart. There are times I wished I had listened and followed that advice more closely. It would have saved me a lot of heartache. It would have saved me from being the pawn in a black heart's game.

We may come away with scars, but those remind us that we survived. We picked up the pieces. We kept going. And because we made it through, we can help someone else make it. Isn't that what it's about anyway? Nothing that happens to us is a surprise to God. He uses every experience we have faced. He will bring someone in our path that needs to know what we have learned. He picks us up. We pick someone else up. That's the real meaning of paying it forward.

Sometimes I wish I wasn't so transparent, but on the other hand, there is a peace that comes from being that true to yourself, and others. You may end up getting hurt. Or you may end up finding love or friendship or respect or your heart's dreams come true. Even the hardest road is bearable when there is someone walking it with you. You have the power to help someone else. The only thing you may be able to do is be there for them. But that can be enough. If we just look a little deeper to the heart of all things, we find we are not that different after all. A shared load is easier to carry. A broken heart is easier to bear when there is someone who understands that pain. We have all been there, or will be there. Be a heart mender.

"Blessed be God, even the Father of our Lord Jesus Christ, the Father of mercies, and the God of all comfort; Who comforteth us in all our tribulation, that we may be able to comfort them which are in any trouble, by the comfort wherewith we ourselves are comforted of God" (II Corinthians 1:3-4 KJV).

I'M ALIVE

Have you ever heard of Newton's third law that says for every action there is an equal and opposite reaction? I have been thinking that for every action there is a reaction. That is a truth I know and I believe. Remember in the Titanic movie, at the end (*spoiler alert!*) after Jack dies and Rose is rescued? It shows pictures of all the things that she did in her life after Jack met his demise in the icy waters. She could have died out there in the Atlantic. She could have slipped into that icy water with Jack and just given up, but she decided to live. Her life was changed. It was changed because she was alive, and realized not only what she had lost but what she still had to live for. Instead of drowning, she decided to fly.

It's funny how when you face devastation in your life, whether its losses of any kind, broken dreams, or broken hearts—one of two things happen: part of you dies, or you realize that you want to live. You become bitter or better. Everything in life is about choices. You may have no control over what is happening, but you have a choice in how you react to it.

I have seen sad times. I have faced dark nights. I have cried more than my fair share of tears. I have been treated cruelly by those I thought were my "friends." I hope that hasn't happened to you, but I think it is something we all face at some time in our life. But you know what? Life is more than the circumstances around you. Life is more than what you possess. It's more than how you are treated. It's more than mind games people play with you or your heart. Life is what *you* make it—not them. You can't control anyone else. You may not be able to control anything. But you do have a choice. When you face the darkest, most painful times in your life you have to make the decision. Do you give up or fight? Do you gather all the strength you have left, and keep going, or just slip into the cold, icy waters and forget it all? We all face that moment. It may happen many times in our life. We have the choice to make. You have to reach the point where your heart says enough is enough. There is more to life than this. You choose to *live*. You're not just going through the motions anymore, but you decide to *live*.

When you decide you want to live life to the fullest, everything in you and around you changes. It did for me. The sun has never been brighter. The breezes have never felt this good before. I know the skies never looked bluer. I am fascinated by the amazing clouds and sunrises and sunsets. I can't get enough of them. (Anyone that knows me can attest to that.) It feels good to run in the sun or rain, wearing tutus, and even crawling through mud. Everything is better because I'm alive. Every breath I take. Every heartbeat. Every laugh. Every day. I can't get enough—of anything. It's because I'm alive. I don't want to take for granted one single breath. Not one heartbeat. Not one laugh. Not one day. Today I opened my eyes again to the gift of another day of life. I'm alive. I can feel it, and I am going to live it with everything I've got. I want to fly. I pray you will too!

"I call heaven and earth to record this day against you, that I have set before you life and death, blessing and cursing, therefore choose life, that both thou and thy seed may live" (Deuteronomy 30:19 KJV).

NOT AFRAID

I'm not afraid of storms. In fact, I've always liked them. I've always welcomed them. I've felt a rush of excitement when one is brewing. Strange, huh? There is something about the intensity and power in a storm that has always mesmerized me. The rain, winds, lightning, and thunder; it's all the chaos that blends together and somehow makes sense. I'm not afraid of the storm.

Even as a small child I can remember as a storm approached my mom warning, "Come inside. Stay away from the windows and doors. No talking on the phone! No taking a shower!" She had all kinds of storm rules. She had more storm warnings than the Weather Channel. But it didn't matter. There I would be. Outside. The rain was pelting me in the face. The wind was whipping my hair. Lightning was flashing all around. Thunder caused the ground to rumble under my feet. It wasn't my smartest moment. But this I know to be true. You can't deny the power of God when you're standing in the storm.

I've never been afraid of the storm. But now I have come face to face with a different kind of storm; a storm of life. Not a cloudburst either, but a full-blown, raging storm. Oh, it's been a doozy. You want to talk rain? How about rain so hard that it blinds you? It hurts to even open your eyes. Wind? The winds are so strong they knock you down every time you can get enough strength to stand back up. Darkness? Don't bother trying to see where you're going because there is no visibility at all. The lightning? Piercing and fierce and constant. And thunder? It's loud and sudden. Just when you're not expecting it—*boom*—there it is again.

If any storm could cause me to fear, this would be it. A sunnier place would have been so much easier. But I have learned important things from this storm that I never would've learned standing in the fair weather.

This storm has caused me to learn to stand when everything around me is out of control. It has taught me to dance in the rain. It taught me that no matter how dark it may be around you, His light will show through. Those flashes of lightning have allowed me to see that there is something ahead if I keep moving. No matter how loud and overpowering the thunder has been, I have learned to hear His still, soft voice in the midst of it all. I would've never learned the depths of His strength, the power of His peace, the undeniable love that would never leave me to face this storm alone. The only way I learned all that was by going through the storm. He has walked behind me, before me, beside me. He has been with me always. The One who controls the natural storm controls this storm, too.

I could've found a way around this storm, but the outcome would've been so different. I wouldn't be who I have become today. So, I thank Him for the storm for it has changed me forever. It hasn't been easy, but it was necessary. And today, because of Him I can still say, I'm not afraid of the storm.

"The LORD is good, a refuge in times of trouble. He cares for those who trust in Him" (Nahum 1:7 NIV).

"The eternal God is your refuge, and underneath are the everlasting arms..." (Deuteronomy 33:27a NIV).

OUT OF THE HALLWAY

I am getting out of this hallway! There are times in our life when we are standing looking down an endless corridor—one that seems to have no end. You know in order to get to the place you need to be, you have to get out of the hallway. But how? What way leads to your destiny? Or what door? Sometimes it's hard to see. The lighting is not the best either so it's hard to tell where the doors are.

The best pace is probably slow, since you are not sure where you are headed, or can't really see clearly what is in your way. You walk ahead, and before long you think you see a door. Upon closer inspection, it is one! You try the knob. Locked. So on you go. Before long another door appears. You try this one, and it opens. You peer inside. There is nothing to interest you in there after all, so you back out and close that door.

Now there seem to be more doors. You look in open ones, hoping to find what you're looking for. You start down the hall again, but stepping in and out of doorways disorients you and you walk right smack into the wall. *Ouch!* That hurt. Life can hit you like that. You think you know where you are going, and then out of nowhere comes that *wham!* You realize you took a wrong turn. Wrong turns can make you unsure of yourself—especially if you take more than one or two. You can lose confidence in your ability to navigate to where you need to be. Being lost is an awful feeling. Being lost and afraid and ashamed to ask for help is an even worse feeling.

If you can get back on course, you still have a chance of finding your way out. But now you step gingerly for fear of getting hurt—or lost again. Pain has a way of making you leery. And more careful. And sometimes distrustful. It can make you a lot of things. Just don't let it make you stop trying. As long as you keep going, you always have a chance of reaching the end.

On you go, and now more and more doors appear. Some are locked. Doors are locked for a reason. They are locked to keep something out or something else in. It pays to respect locked doors. I have banged my fists on some doors trying to get them to open until all I was left with was sore knuckles and a heavy heart. I really wanted in those doors. But they were locked for a reason. Good reasons. It's best to learn to respect that. You will save yourself a lot of heartache and regrets. It may be locked so you'll take a different way.

Then there are the open doors. There are always open doors if you look for them. They can blend into their surroundings at times, but if you look hard, they are there. You begin investigating some. There are doorways to all kinds of things. Some are full of good things. Some are full of things you should avoid. Some you pass without hesitation. But after awhile some that you should keep moving past seem to draw you. You shouldn't go through them, but in moments of weakness, good judgment goes out the window, or, in this case, out the door.

Once you go through some doors it's hard to find your way back out of them. Some people go through doors, never to find their way out again. They are lost in plain view. They never reach their destination. I don't want that to be me. I want to do what I was put here to do. I want to be who I was meant to be. I want to find the purpose for my life. That means you need to go through the right doors.

Learning to understand which ones to enter and which ones

to walk or run away from can be hard. It takes wisdom and experience to recognize the difference. Sometimes that comes from going through wrong ones. You learn (sometimes the hard way) that not every door is meant for you. If only there were a magic set of keys that opened all the right doors. No such luck. So we walk each step by faith, trusting that we are headed down the right hallway and to the right doors. We all open wrong doors now and then. It could be from curiosity or just being human. You can get really tired hanging out in the hallway, but we often don't have a lot of choice. The best thing to do is keep moving, and believing that the right door is going to open for us at the right time.

God knows all that we go through. The right doors. The wrong doors. He knows how tired we get of searching. He knows we are weary and worn. He is never far, although at times He seems silent. He is waiting for us to find the place we should be. He could just put us there in front of the door, but what would we learn from that? You never appreciate something that costs you nothing. You find the real value of something when you know its worth. When you realize what you gave to find it, you hold it as you would a treasure.

Wait. Did you hear that? It sounded like a door opening just ahead. I hear that assuring Voice that has led me to this place. "I know all the things you do, and I have opened a door for you that no one can close. You have little strength, yet you obeyed my word and did not deny me" (Revelations 3:8 NIV). If we are faithful to God, He will get us where we need to be. To the right doors and out of the hallway.

We may be weakened from the searching, but just across the threshold all will make sense, and we will find a place to rest and understand. We'll be out of the hallway—finally.

Duct Tape Can Fix That

Get me the duct tape. Duct tape can fix anything. Have you seen all the colors and designs it comes in? I needed some for a school project and spent a *long* time looking at them. It was hard to decide what to choose. It was a lot easier when the only choice was gray. Duct tape is kind of amazing. It is so strong, and man, can it stick. If you need something stuck together, get the duct tape.

Have you ever gone through things in your life that left you feeling so discouraged? Your heart is tired and worn. Your shoulders have carried a heavy load too long. Life is exhausting. It seems like all that's left is hanging threads and pieces. How will you ever keep it together when it's falling apart? I was going through a really hard time, and I told a friend at church one night that the only thing holding me together was Jesus and spiritual duct tape. I wasn't kidding either.

As great as duct tape is, it is not a cure-all. It will however hold things together securely for a good while. Sometimes that's all you need. Something to keep you in one piece until you get where you need to be. You need to get to the place where you can find forgiveness for your failures, healing for your hurt, and strength for your weaknesses.

You may ask what's the use in trying? You're never going to win. You'll never overcome. You'll never measure up. Yesterday the devil reminded me who I'm not. But today, I'm reminding him who I *am*. I may not measure up to some people's standards.

You know what? I don't care anymore. I am not defined by who they think I am. I am defined by who He says I am. We may be pressed down by troubles, perplexed by situations, and knocked down. But we are not crushed, destroyed, or abandoned by God. No, we're not quitters. We're survivors. Sometimes we fall. We get back up. Maybe you have fallen. Don't stay down. There is strength and mercy all around.

We have come so far. We still have a long way to go, but we're going. We're going to make it. Some days all that's holding us together is mercy, grace, and spiritual duct tape. But that's enough. Nothing sticks like duct tape, except maybe gorilla glue, but that's another story for another day.

"Who shall separate us from the love of Christ? Shall tribulation, or distress, or persecution, or famine, or nakedness, or peril, or sword? As it is written, For thy sake we are killed all the day long; we are accounted as sheep for the slaughter. Nay, in all these things we are more than conquerors through him that loved us. For I am persuaded, that neither death, nor life, nor angels, nor principalities, nor powers, nor things present, nor things to come, Nor height, nor depth, nor any other creature, shall be able to separate us from the love of God, which is in Christ Jesus our Lord" (Romans 8:35-39 KJV).

LIFE IS LIKE A BOX OF CHOCOLATES

"Life is like a box of chocolates. You never know what you're going to get." That is so true. We can make plans, but that doesn't mean we're going to get what we want. It's like when you get that big, heart-shaped box of chocolates. You know what I'm talking about. It makes you feel so special just looking it.

Later (when you're alone) you open it up and dive in. You pick your favorites first. Those are the ones you know. Sometimes it's hard to tell what each piece is. If you're lucky, it'll come with a paper telling what piece is in each space. If it doesn't, you have options. You can take a bite out of every piece. It is, after all, your box of candy. If you're more conservative, you can poke a little hole in the bottom of each piece so you can see what's inside. If you don't like it—pitch it. (Unless I am really desperate, the dark chocolates are the first to get tossed.)

I realized that God has given us a "box" of life's assortments. Some things in our life are like those favorite pieces of chocolates. My favorite piece happens to be the coconut. Or maybe the pecan and caramels. Hard to decide sometimes. Either way, I know them and I'm comfortable with them because I know what I am getting. Some things in our life boxes are easy to identify. They are the ones that we can see what God is doing and where He is leading. They are the "caramels" in our lives.

They are the easiest to pick out because we have learned what they look like. We know what to expect. There is no surprise when your teeth sink into the chocolate. Then there are the pieces that we are not as familiar with. Sometimes we have to study them, looking them over till we can figure out that God is giving us something deeper and richer than we're used to. We learned to appreciate something that we haven't tried before.

Then there are the pieces that we have absolutely no idea what they are. We pick them up. Turn them over and over and over. We can see them, but we don't feel comfortable about them. We haven't had to handle anything that was like this before. We ask, "What are You giving me Lord? There's nothing about this that looks even vaguely familiar." Some things we face make no sense. There just aren't any answers to some questions. If it was just always easy to understand the things we are going through, and the reasons we have to face them. But that's not how it works. We walk by faith. We have to trust when we don't understand. Believe even when we can't see. So we decide, we're not going to live by what we see. We're not going to live by what we feel. We may not recognize what God is giving us. But He is giving it for a good reason.

It may not be the one we would've chosen at first glance in the box, but we just know that it we will like it better than what we may have picked. God doesn't make mistakes. He always gives His best to those that love Him. We may not see and understand the reason now, but farther down the road of life it will make sense. We will see the way He worked things out for the best. And just like that favorite piece of chocolate, we can know His plan for our life will be sweet and rich, and we are so going to enjoy it!

"Trust in the LORD with all thine heart; and lean not unto thine own understanding. In all thy ways acknowledge him, and he shall direct thy paths" (Proverbs 3:5-6 KJV).

A Walk On The Desperate Side

Desperation. You find yourself in an emotional state of despair so deep; in a situation that looks so hopeless that you are driven to resort to extreme measures in order to survive. That is one heavy analogy of a complicated feeling. I have always tried to look for that silver-lined cloud... for the sun coming up in the morning. I have been known to burst into a rendition of that

Annie tune on more than one occasion. I look for the good in the bad. I look for the light in the dark. I look for the hope in the hopeless. Even the most optimistic person will tell you, we all have down moments or even down days. It's not about never feeling down. It's about not *staying* down.

So let me be perfectly honest. There have been times when I have been...desperate. Have you ever seen a turtle on its back? Now that is a desperate situation. He knows he is in trouble, and helpless to some extent. At least it can appear that way.

Desperation is not a calm, peaceful emotion. Oh no. When we talk about desperation, we are talking about a top participant in the X Games of Emotions. Intense. Fierce. Dire. When you reach the point of desperation, all rationality heads for points unknown. Being rational would mean not just "reasoning" out the best choices, but possessing and making logical decisions based on careful consideration. Desperate people do desperate things. When you reach the point of desperation, all common sense and sensibilities takes a hiatus. Under normal circumstances, you accept some things may not go right or work out as you planned. But in desperate times, that just doesn't matter. You are that turtle on its back *desperately* trying to get back up.

There comes a day when we all end up in Desperation Town. If you have never been there yet, hold on, your day of visitation is coming. Hopefully, it will be a short introduction and you can get away quickly. Not everyone escapes that easy. No, sometimes desperation takes more than it rightfully deserves.

It can take your self-respect. It can take your dignity. It can take your self-worth. In moments of desperation you may do, say and act in ways you never would have if you weren't desperate. It's not always a good feeling. It can be an overwhelming feeling that pushes a person to a point of breaking.

There have been times when I have been so desperate, wanted something *so* badly that I was willing to compromise who I was and what I believed because I thought I needed it that much. It wasn't my proudest, or happiest moment, because that kind of desperation always leaves you—wanting. Have you ever been in that place? It's a place that can hurt just by remembering being there.

But desperation has a good side. Being desperate can push you to go after what you need or want. Or it can take you to the place where you realize you are so far from where you should be or could be or need to be, that you hit the cold, hard ground of reality and wake up. It is the harder push of the struggling turtle's legs and neck to get his world, and shell, turned right.

Desperation is actually a necessary ingredient in the big scheme of things. Because in desperation, you learn. You learn to accept. You learn to create. You learn to fight. You learn to hope. When you reach that point of desperation where you think you would be willing to do anything to get what you want, you learn who you really are. You learn to change—not the situation but you.

Desperate people do desperate things. There are no desperate situations, only desperate people. I have been one of them. For most of us, it isn't a permanent state, just a momentary blip (or a couple of momentary blips) on our emotional radar screen. You need to hold on to the hope. Life is not always what it should have been, or could have been, it just is what it is. We should remember that thought every day, and make the most of it. George Washington said, "We must never despair; our situation has been compromising before, and it has changed for the better; so I trust it will again." When it looks like the darkest defeat is coming, victory may be just ahead. There is dark before the dawn. The sun will come up. Don't despair. I hope you never have to find yourself in the state of Desperation, but if you must, may your visit there be quick and as pain free as possible.

"In my desperation I prayed, and the LORD listened; he saved me from all my troubles. For the angel of the LORD is a guard; he surrounds and defends all who fear him" (Psalms 34:6-7 NLT).

Rip Tide

Have you ever been caught in a riptide? Talk about a scary experience. What makes them so frightening is that they just seem to come out of nowhere. You can be swimming along, enjoying the waves and sun, and then you are gripped by something you have no strength to fight. The first thing they tell you *not* to do is the first thing you do. Panic. It seems logical to me. I mean—have you seen the size of the ocean? Suddenly you are being swept out to the vastness by a power you can't compete with. I was caught in a riptide once. It was terrifying, but I made it out—evidently. I have been caught in "spiritual" riptides, and believe me, some of them have been harder to escape than the watery kind. Even in life, we can get lulled into a false sense of security or just not be paying enough attention to little signs around us.

It came out of nowhere. I was just floating along happily, when something latched on and began pulling me under and away from the safety where I was. You realize immediately that you are in trouble. The first thing you do is start fighting with all you have to get free. Survivor mode kicks in. But the funny thing about riptides is that, the harder you fight, the worse your situation gets. A drowning happens because the swimmer exhausts himself fighting the force. You don't want to go under. You don't want to be pulled away. You don't want to die. I remember when I was caught in the real riptide, after the initial shock, even though I was afraid, reasoning began to kick back in. I remember hearing that if you swim parallel to the shore, you can gradually swim out of the currents' control. So that's what I did. It took a while and was still hard and exhausting work.

Now my "spiritual" riptide was a little different. Oh, I was in danger. I was going under. But how could I escape the grasp pulling me? Where do you find the strength to fight when your strength is all but gone? How do you calm yourself enough in the midst of the fight to listen to the still small voice whispering His promise and peace to your heart?

Another tidbit of advice is if you keep your feet on the bottom the current can't get a hold of you as easily. Hmmm? That's good advice in life. Keep your feet *firmly* on the foundation, then whatever sweeps around you has less chance of catching you off balance. Sometimes you are in over your head. You can't always stay near the shore. Who would want to anyway? When you find yourself caught in things you can't control, things you have no strength to fight—just stop fighting. Stop trying to control everything. It's exhausting. It is pointless, too. There are some things in life you can't fight and win. But He can.

So at my weakest, lowest point—I gave up. Not the battle just the fighting. I turned it over to the One that controls the waves and the winds—at the shore and in my life. I quit struggling against everything and just started heading back to where I know safety is. Slowly. I know I'm not alone. The grip has loosened. Even in my weakness, if I just keep my eyes on Him, I find the strength to go where He is taking me. If you are caught, don't fight it. Look to the One that has the strength to carry you through. Tread water, but don't stop moving. You are closer to the shore than you think.

"When you pass through the waters, I will be with you; and when you pass through the rivers, they will not sweep over you. When you walk through the fire, you will not be burned; the flames will not set you ablaze" (Isaiah 43:2 NIV).

But I Love Those Flip Flops

I have been known to spend a lot of time in swampy areas. I am quite comfortable tramping around the bulrushes and cattails, often in old flip flops that soon are covered in mud and muck. On more than one occasion I have tossed them in the washer and hoped they came out in one piece for another adventure.

I looked at my old flip-flops the other day and asked myself,

"Are you *seriously* going out in public in those?" Hey! I love those flip-flops. In fact it was love at first sight. They were so cute and fit perfectly. They didn't even need broke in. They just belonged on me. We travelled many miles together. But something happened to them. They're not looking so hot anymore. They still make me feel good. They're comfortable. They're familiar. They're my favorite. But they have seen their better days. How did this happen?

Once again God turns a simple thing in my life into an object lesson. He tells me it's time to let go of something that I have held on to for far too long. I know what He wants me to do, but I don't want to. But this feels safe I tell Him. This feels secure. This feels comfortable. He says it's time. I say don't think I'm ready. He tells me their purpose has been served. You can only kiss so many frogs before you realize they are not turning into princes.

Just like my flip-flops that helped me through many summer days need replaced, it's time to move on to something new. You mean out of my comfort zone?? They say when God wants you to grow, he puts you in uncomfortable places. That's where you feel like a frog out of the pond on a hot day.

He says I'm not alone. I say I don't know what is ahead. He says He does, and it is *so* much better than what I have now. He is saving the best for me. But if I want it I will have to get rid of the old that's holding me back. (You realize we are not really talking about flip-flops now, right?) I can't hold the old and new. I can't wear old flip-flops and new flip-flops at the same time. It just won't work. The old just can't support me where I'm going. They're not strong enough. I'm going to need strong.

The unknown is always scary territory. But what is even scarier is to hold on to the familiar and ordinary when following where God is leading will bring the extraordinary.

I know that's true. Who wants to settle for mediocre? We

should never settle at all. God has promised His best, and that is definitely what I want. So the first thing to do is begin. How? One step at a time. The first step is always the hardest, but then each step will get easier. Maybe you have some old "flip flops" that you need to put off. It's time to step in faith. We know in whom we have believed. Walk on and trust and hope in Him. I will be wearing new flip-flops though. I want to put my best foot forward. God's best is waiting down the road

"Since you have heard about Jesus and have learned the truth that comes from him, throw off your old sinful nature and your former way of life, which is corrupted by lust and deception. Instead, let the Spirit renew your thoughts and attitudes. Put on your new nature, created to be like God—truly righteous and holy" (Ephesians 4:21-24 NLT).

Ready To Roll-er Coaster

When you love roller coasters, it's all about the ride. You love that feeling that is different from any other kind of rush. You wait in endless lines, with anticipation, long before the ride ever starts. Funny how the time waiting in line never equals the actual ride. It's always short in comparison. But it doesn't matter. You don't mind waiting—the ride will be worth it. My life has become that roller coaster ride. I was looking forward to the thrill of a new adventure. It can make you a little nervous. The unknown can be scary ride. I had to board my seat. Which one? Front row! Why not?

I rode up the first hill with excitement and trepidation. I was leaving the familiar where my feet had been firmly planted on the ground and now was rapidly advancing into unknown territory. There is that place as you crest the hill when you're able to take one last deep breath before the plunge ahead. Then it starts. The face-forward drop. Then there is all the jerking back and forth and side to side. Restraints or not, you are banged around. The wind makes it hard to breathe and the fact that you realize you're screaming at the top of your lungs doesn't help. One moment you are screaming, then the feelings change and you are laughing.

Coasters have a way of doing that to you. So does life. The highs hills and lows dips. Accelerating speeds. The tunnel-like places with a darkness you can feel. Just when I am lulled into thinking the ride is slowing down we begin another ascent. Everything looks so different from this ride, too. Things that I used to know so well I can hardly recognize at these speeds and heights.

As much as I love the thrill it brings, this ride I'm on is getting old. Up and down, side turns and twists, loops and corkscrews, and the all the upside down turns are making me nauseous. I am beat up from being knocked back and forth. While I love the adrenaline rush, I just want to be able to relax.

I want to put my feet on something solid for a change. I want my world to stop rocking and flipping.

I hate riding rollercoasters alone. There is another dark tunnel ahead. I reach out in the darkness desperately grasping for something to hold on to. Then I feel His hand. I'm not alone. Strong arms hold me securely—steadying me from the constant tossing. The same God that stilled the stormy seas stills my racing heart. His voice calms my fears of not knowing what is ahead. I know this ride won't last forever. It will soon end. It is in the journey that we learn.

I have learned so many things on this ride. I am departing it a different person than I was when I started it. Braver. Wiser. Kinder. Stronger. It has been a wild ride. It has been scary, breathtaking, and thrilling all at the same time. But, Lord, how about we head over to the Merry-Go-Round (or Tunnel Of Love) next?

"We are troubled on every side, yet not distressed; we are perplexed, but not in despair; Persecuted, but not forsaken; cast down, but not destroyed" (II Corinthians 4:8-9 KJV).

"Peace I leave with you, my peace I give unto you: not as the world giveth, give I unto you. Let not your heart be troubled, neither let it be afraid" (John 14:27 KJV).

WHEN LIFE GIVES YOU LEMONS

I know summer is unofficially over after Labor Day, but have you been to my lemonade stand? Oh, yeah. I have taken up selling lemonade. You know how they say "when life gives you lemons..."? There were times I wondered if I had a sign on my back that said to drop all lemons off here. When it rains, it pours, and in my case it rained lemons. And what do you do when you accumulate an overabundance of lemons? Hmmm.

I don't remember anyone ever saying life would be easy. I guess they just didn't explain how hard it could be sometimes. There are the big things we face. Those things that crush our heart and sometimes shake the very ground we're standing on. Things that like lemon in big doses, can leave you sour. There are the disappointments and discouragements that come our way. There are things that didn't turn out like you expected or wanted. There were things that you were counting on that fell through. There were people that you trusted who let you down. Those are the things that can leave you bitter like small tastes of lemons. Not enough to make you pucker but just enough to leave you with a bad taste in your mouth and in your heart. And who wants a sour and bitter heart? Not me. They tell you "when life gives you lemons, make lemonade. But just like it takes sun and rain to make a rainbow, it takes more than lemons to make lemonade. You need water, but more importantly, you need sugar. It doesn't work without the sugar. It can be refreshing, but it just doesn't taste the same. But where can you find the sweetness in life when it feels like you have taken up residence under the lemon tree? Look for the good. It's there. It may seem small, but all you need is a place to start. "All things work together for good..." You may not see it now, but down the road you will. Have hope. "Why am I discouraged? Why is my heart so sad? I will put my hope in God..." (Psalms43:5a NIV). There are so many things we can't do or fix or change, but He can. You can hope. Smile. Tell yourself you are going to be happy. You know what? You will be. If you are collecting lemons, do what you have to do. Start squeezing. If you need some help, call me, as I am an expert at making lemonade.

ARE WE THERE YET, LORD?

Are we there yet, Lord? Can you tell me where I'm headed? Couldn't you let me see a little farther than what's right in front of my face? I know there are good things ahead and I want to get to where they are. Couldn't You just pick me up and drop into the place where You're leading me to. That would be *so* much easier. No? I have to get there step by step. Day by day.

Why can't I be there yet, Lord? Oh, I'm not quite ready for what's coming, is that it? So I have new things to learn that will help me. I have some battles to fight (yuck) that will give me skills I'm going to need for what's ahead. I need to buff off some rough edges. Sounds like a lot of work. I'd rather do this the easy way. But easy is not always better. And if given the choice, I'd rather have the best not just better.

I have put my life in Your hands. So I guess that means going with Your plan. A few more spins on the Potter's wheel. A few more stitches in the tapestry of my life. You are the Master of everything You do, so I know if You have control of my life, it will end up a masterpiece.

I am learning what it means to walk by faith, not sight. I'd like to see where this road is leading, but I can't see that far ahead. So for now, I am just going to enjoy the journey.

I want to smell some roses along the way. Even they have to unfold a petal at a time to reach their perfection. I want to skip a few stones on the pond of life. I will follow one day at a time, knowing that before long I will get there and it will all make sense. Are we there yet, Lord?

"But He knows the way I take; When He has tried me, I shall come forth as gold. My foot has held fast to His path; I have kept His way and not turned aside" (Job 23:10-11 NASB).

You Are Worth More

How much is a sparrow worth? They are so common. You see them everywhere. The Bible says five sparrows sell for two cents. That's cheap. Today for a while I felt like a sparrow. We live in a world that cheapens everything. Even the worth of people. But I am realizing I am worth more than many sparrows. No one is putting a cheap price tag on me again. I am a child of the

King—that makes me a princess. The King is not pleased when someone disrespects, dishonors, and misleads his daughter.

Am I at fault? Yes. Sometimes princesses can be head-strong and rebellious. When you always try to see good in people, it makes you too trusting to believe that there are people in the world that would take advantage of someone's honest, true feelings. Can someone be that deceiving or just unfeeling? It's sad, but there are people like that on the world. They don't see you as another person. But it was really my mistake. Really. Back in the garden the serpent tempted Eve. We are still tempted and beguiled. I was tempted and I trusted. But we need to remember that decent people don't treat people as objects. We should forgive and move on, but don't forget the lesson you learned.

I wish to God that I could say I was always strong enough to withstand what the devil tries to lead me into—always smart enough to see through the motives and desires. But there are still wolves in sheep's clothing. They know their weaknesses and admit them. That makes them seem sincere. They could change from being a wolf. God can change anyone. They have to want to change and actually do just that. Change.

No matter if they change or not, we need to. Don't let anyone make you feel that you're not enough. That you don't matter. That you're not worth it. Don't be led down paths that you never would journey down if you didn't let your heart lead instead of your head.

One day a sparrow hit the windshield of my car. I was on a country road so I stopped to see if it was all right. It wasn't. I felt sad looking at its small, lifeless body lying there in the gravel. I had been having a rough time and seeing that tiny sparrow was a catalyst to my already frazzled emotions. Tears began to stream down my cheeks. It seemed so pointless. The sparrow had just lost its life. It was flying free and then suddenly it was dead and alone. It was like its life hadn't mattered. Who would care about it lying there along the road? I was feeling worthless and alone myself at that moment. Then I remembered how the Bible says God attends the funeral of every sparrow. Every sparrow. Someone cared. Even for the small, seemingly insignificant bird. The Creator of the universe and all things takes time to attend each funeral. He knows what is going on, and He has it all in control. I found comfort in that.

I pray you can feel that same comfort. You may not think you are worth much. "You are of more value than many sparrows" says the God that attends every sparrow's funeral. *He* knows your true value. What a picture of a loving God. You are bought with a price—His precious blood. The sacrifice He made wasn't cheap and neither are you. Hold up your head. You matter. Don't ever let anyone make you feel otherwise. We should pray for those that have hurt and misused us. We are all human. We all make mistakes, and until we make it home will battle this flesh every day. I am putting a big "Not For Sale" sign on my life (and a "Not For Giveaway Either" sign). We are worth more. My Father said so.

"Are not two sparrows sold for a penny? Yet not one of them will fall to the ground outside your Father's care. And even the very hairs of your head are all numbered. So don't be afraid; you are worth more than many sparrows" (Matthew 10:29-31 NIV).

A Song In The Night

I woke up one morning to the sound of a bird singing loudly outside my window. Not an odd occurrence. Most of my mornings start out being serenaded by song. The difference today was that it was 3:00 AM. Through squinting eyes I peered at the red numbers on my clock to check the time. Wait a minute. This is a little early for morning wake-up songs. I laid there listening in amazement. I guess the shock of surprise tweeting in the middle of the night was enough to pull me to "fully-awake" mode. The lone bird started out with a soft and sweet song to welcome the day. It was a call to other birds to respond. "Birds of a feather flock together" and all that. And yet not one other bird joined or responded. The tweeting slowly changed.

The call became almost like a question waiting for a reply. But no reply came. I think the bird realized it was alone, in the dark, with no one to answer. It was a sad, lonely song. At that point it turned different yet again. It turned from lost and desolate to frantic and desperate. I know that feeling. When you feel like the walls are caving in and no one is there to help hold them up, or at least dig you out from the rubble. I looked at the clock. I had been listening for almost 45 minutes to that little bird's distress calls. Then came a soft, pathetic cry of something scared and alone. I recognized that cry. There have been nights when I felt like I was so alone that no one could hear my cries. It is a scary and lonely place to be.

But what makes a bird sing at night anyway? Nightingales often do. Their name explains that. But why would other birds do it? Sometimes changes in lighting confuses the bird as to what time of day it actually is. They can be startled awake by some noise, and just like me, unsure of what is happening. Birds aren't the only ones that experience rude awakenings. That happens to us, too.

Do you know how you think everything is finally coming together? It looks like everything is falling in place. But then from our peaceful state we are shaken. We never even saw it coming, but we are suddenly jarred to our core. That bird, for whatever reason was jolted out of its sleep and dreams to find itself awake and disoriented. It is like being awakened from a dead sleep. You shake yourself, sometimes literally, to help get enough consciousness to grasp what's happening.

The only problem is that sometimes there is no rhyme or reason for the things we face. Bad things do happen to good people. And when they do, we are like that bird. We first figure that we will face the unexpected with a song and a brave face. But when we feel our faith waiver, we put out a call for help.

There will be times that for whatever the reason, no one comes. Reality can be a cold, harsh place to face. Especially alone. It can make you act in ways you might not normally act. Do things you might not normally do. You just want someone to answer your call; someone to let you know everything will be all right. That you will be all right. You know it may not be immediately. It is still pretty dark. You just want someone to tell you morning is coming, and things will again look familiar and make sense. You are not alone.

I can tell you that fact from personal experience. You. Are. Not. Alone. Even when there is no one to answer your call, no one to hold you, no one to wipe the tears—you are not alone. Hard places teach us lessons that we wouldn't learn at any other time. It has been in the dark lonely nights that I have found there is One that is always near. You don't have to face the dark alone.

He will always answer your call. "He will cover you with his feathers, and under his wings you will find refuge; his faithfulness will be your shield and rampart" (Psalm 91:4 NIV). Or in other words, He will be your place of shelter, the One that blocks things thrown at you, and your defensive wall. I will take all that.

He cares for the sparrows. Remember that God attends the funeral of every sparrow. Doesn't He care more for us? Oh, I am sure He does. In fact, I know He does. He has covered me. He has comforted me. He has protected me more times than I could count—even at times from myself.

You are not alone. I know it can feel that way. But you are

not alone. Even if it is faint and weak, let your song rise up in the darkness. There may be someone else that is listening for a reason to keep hoping. Let them know they are not alone either. It may be dark. It may be scary. But, you are not alone. His eye is on the sparrow, and I know He watches me.

"I lie awake, lonely as a solitary bird on the roof" (Psalm 102:7 NLT).

"By day the LORD directs his love, at night his song is with me—a prayer to the God of my life" (Psalm 42:8 NIV).

"I will remember my song in the night: I will meditate with my heart, and my spirit ponders: (Psalm 77:6 NASV).

"Yet they don't ask, Where is God my Creator, the one that gives songs in the night? (Job 35:10 NLT)

LOST CAUSES

A lost cause? Have you ever felt like one? No matter what you do or how hard you try things just don't work out. You feel like the poster child for Murphy's Law. The poster would read "If anything can go wrong..." and there would be your smiling. Well, probably not smiling face.

You are not alone. Haven't we all at one time or another had a day (or few days or even years) when it seemed like all was lost? We have felt like the stray cat nobody wanted.

I've always had a soft spot in my heart for Charlie Brown. Costume full of holes and the bag of rocks. Empty mailbox on Valentine's Day. Sad Christmas tree. Seems like another lost cause. Who can relate?

The world is full of lost causes. Lost causes that just need a little bit of hope. They're looking down, but want desperately for someone to give them a reason to look up. You could be that someone. You could be the difference in the direction they are setting their sights. From low to on High. From hopeless to the Hope. Jesus came for the lost causes. For the outcasts. The left outs. The nobodies. The strays. He came for the unlovable, the lonely, the weak, and the unwanted. And you know what? "Such were some of you...." That's right. We have all been lost causes. One day I was the outcast, the lonely, and the hopeless. But you know what? I found a reason to hope.

He picked me up and now I want to be a hand that helps lift someone else up. I want to be the voice that speaks words of peace to the troubled heart. Show a love that changes a life. Yes, the world is full of lost causes desperately looking for a light in their darkness. Be that light. Seems like a hopeless cause. Not really. We can do it. Let's start today and make our world a little less lost—and a lot more hopeful. One lost cause at a time.

"For all have sinned, and come short of the glory of God" (Romans 3:23 KJV).

"And such were some of you: but ye are washed, but ye are sanctified, but ye are justified in the name of the Lord Jesus, and by the Spirit of our God" (I Corinthians 6:11 KJV).

I Know Why The Caged Bird Sings

"I know why the caged bird sings." I know, because I used to be one. When the door opened, I crept out into what was a strange and scary place. Finding your way is no easy task when you are used to being confined. Suddenly there's this huge world. So much in it makes no sense. But I know why the caged bird sings.

The past few years have been full of changes. In fact, change and I are now best friends, because it is an ever-present force in my life. Change can be scary. Change can be painful. But the best kind of change is the necessary. Not that it is easy. Oh no no no no no. It is the hardest, because it is the change that has to be done if you are going to succeed and survive.

I look back at who I used to be. I don't even recognize that little bird.

"I know why the caged bird sings."

When I was in the cage, I sang. Because I was happy? No, but because I had hope. Hope is a powerful thing. It kept me alive then. It keeps me going now. No matter how many times things don't turn out the way I expect, and I feel like giving up, that voice deep inside says "No! it's not over yet. *Get back up!*" And I do.

I summon what seems like the last strength I have because I know that until I give up there is always a chance. As long as I can keep going there is always hope.

This world outside the cage is a baffling place full of confusing people. But the same world can be magical and beautiful, and there are people that help you find places you never knew existed.

Every day may not balance out, but the overall story will eventually. I sang in the cage because I had hope, and I still do. But now I sing because I am happy. Things may not be perfect but what is? As long as we have breath, there is reason to hope and sing.

"You will be secure, because there is hope; you will look about you and take your rest in safety" (Job 11:18 NIV).

Don't Touch My Rose-Colored Glasses

Don't touch my rose-colored glasses! Yes, I have been accused of wearing them on more than one occasion. That's OK. I like how my world looks through them. Pink. Rosy. Happy. There's enough dark stuff in the world. I want to see the roses. I want to see the sun and the rainbows.

Don't touch my glasses. You don't just get these glasses for nothing. No. They come at a cost.

It's all about choices. We get to choose how we are going to handle what life throws at us, or in some cases, drops on us. Life has even crushed me on more than one occasion. I had to make a choice every time. Stay down. Give up. Quit. Or *get up*. Move on. Put on the glasses. Yes. I have earned the right to wear these glasses. When things were bad, I have learned to find the good. When all hope looked lost, I found enough faith to still believe. I may not see where God is taking me, but oh, I am hanging on to Him with everything I have. I refuse to accept that He could fail me. That is just impossible. He can never fail. He will never leave me or forsake me. Ever. So I can wear the glasses. No matter how things may look...it doesn't matter. God has got this. He's got me and you in His hands. So put on the glasses. Choose to see with faith. Everything looks better when you look with faith. Even in the dark—even when it's hard—even when you have no reason to, keep the faith. Put on the glasses. You won't be ignoring the obvious. No, you will be accepting the inevitable. God is going to do it. So while I am waiting, you can bet I will be wearing the rose-colored glasses. Pink has always been my color.

"So we don't look at the troubles we can see now; rather, we fix our gaze on things that cannot be seen. For the things we see now will soon be gone, but the things we cannot see will last forever" (II Corinthians 4:18 NLT).

I Wish I Were A Butterfly

I would love to be a butterfly. They are beautiful and free. But they don't start out that way. Oh no. They have a much more humble beginning. A caterpillar. Sometimes it's a grubby looking one, too. But everything has a God-given purpose. Caterpillars crawl around. Slowly. They are low on the food chain. They should watch their back, because there is always something bigger out to get them. Something may pick them up, and sometimes squash them. (I have worked with preschoolers, remember?) Caterpillars are pretty defenseless out there in the big world.

Their one goal is to eat enough so they can be prepared to spin their cocoon. That is where the magic happens. The change. Oh, I could relate to that caterpillar. Feeling small. Grubby. Picked on. But there is always a purpose for what we go through. We go through times where it seems like we are moving so slow. There are struggles. Things to overcome. Things to move past. Then you have to spin the cocoon. Looks easy if you watch. But that caterpillar is putting everything it has into that cocoon. I know what that's like, too.

You would think when you make it to the cocoon, it would be restful. No. It is a place of change. Wings growing. Body changing. Tight space. I know what the cocoon feels like. Change is not always easy. Sometimes it hurts. But it brings us to a new beginning. A metamorphosis. Then comes the day when the caterpillar emerges from the cocoon. Another struggle. Think about it. All restricted in that small space. But it knows. Yes, it knows. Freedom is within reach if it can just get out of the cocoon. So it fights for its very life. It wants to get the new creature that it has been becoming out. It has been dark and dry and lonely. There is light out there. It will be able to breathe and fly. It's a fact that unless it fights its way out, it will be crippled and never have the ability to fly. It would be a butterfly with beautiful wings that it could never use. It's the struggle that strengthens the wings. So it pushes on until it breaks out completely. Can you imagine what it feels like the first time it spreads its wings? I know exactly what it feels like.

To feel the sun on your face. To fly uninhibited for the first time. To be changed and free. Yes, God uses all we go through to change us and to make us something more than we ever imagined we could be. He wants to help us reach far beyond what we started out as into something spectacular in His hands. So wherever you are in your struggle today, know that He is changing you into something far better than you ever imagined. It will be worth the fight. The caterpillar will become a butterfly. You will fly.

"Therefore if any man be in Christ, he is a new creature: old things are passed away; behold, all things are become new" (II Corinthians 5:17 KJV).

GYPSY (NOT GANGNAM) STYLE

I think deep down inside, I have always had a gypsy spirit. I am sure those around me have seen her coming out the past couple years. That must be where the desire for bright colored, flowing skirts is coming from. Granted, a lot of mine are tutus, but hey—they are certainly flamboyant to say the least.

Some clothes just make you feel like dancing. Gypsies are known dancers. King David was a well-known dancer. He danced with all his might. I have found myself breaking into dances, too. There is something about dancing that makes you feel happy.

Another gypsy quality is they love being free. I guess it's not just the tutus that make me feel happy either. I like to put together outfits that make me feel good. I have developed my own sense of style, and I finally feel comfortable in my skin. We all need to reach a place where we accept and love who we are. Until you truly love yourself, how can you expect anyone else to?

The great part about learning to love yourself, is that you learn your self-worth. That isn't conceit. It means you accept who you are, and can stop doubting that you are enough. You realize that you have value. You can stop being afraid to be who you are. It shouldn't matter if someone doesn't see all the great things about you. But it does. Oh, it can hurt like a stab to the heart. (And I am sure any self-respecting gypsy knows all about the perils of playing with knives—and taking chances on love.)

You get hurt sometimes. Really hurt. But don't let it change how you see yourself.

I have gone through that very thing. You know what this gypsy did? I had a good cry, tossed some stuff around, hit a few things in the room, picked myself up, brushed off my skirt, and moved on. That's gypsy style, right? Of course it is. That's the gypsy's adaptability to change. I definitely have that trait. (That ability alone makes me prime gypsy material.) I guess that comes from always traveling around, not putting roots down for too long in any one place. I am beginning to relate to that, too. That may seem like a sad thing, but it's really not. There is freedom in knowing you are not tied forever to one place. After all, this body is just the traveling caravan for our soul. We should never get so grounded in one spot we miss the possibility of the adventure that is calling us. Or it may be the voice of destiny. Either way. You better answer it. When it all comes down to it, we are all only pilgrims in this world anyway. We are all just passing through. Maybe the Pilgrims were actually gypsies. Now there's a thought.

"You have turned my mourning into joyful dancing. You have taken away my clothes of mourning and clothed me with joy" (Psalm 30:11 NLT).

"But they were looking for a better place, a heavenly homeland. That is why God is not ashamed to be called their God, for he has prepared a city for them" (Hebrews 11:16 NLT).

READY FOR SPRING

I was ready for winter to be over—*yesterday*. I am tired of the cold and the snow and the gray skies. I want to see the sun in the blue sky and feel the warm breeze. I want to see the green grass and leaves and flowers everywhere. I want to wear flip-flops. I know there is a purpose in every season in nature—just like in life. I know there is a purpose to winter.

There are benefits from that deep, winter rest. It is a time when you are dormant, so to speak. When it gives things in your life a chance to heal and regain all the energy you have had to use getting through the other seasons you have come through.

But then you feel that restlessness. There is a stirring. Something is waking up. Deep inside. The break you have been on has accomplished what is was supposed to. And just in time. Your heart that was hardened by hurts and disappointments and life in general has begun to soften just as the ground does in spring.

Now a new strength and purpose is beginning to push through. It's not easy coming out of the winter. But you can do it because, though it was cold and long, the winter gave you things you needed. Time. Rest. It is a chance to see things in the stark reality of the winter's background. It gives you a new perspective.

I am ready for spring now. I am ready to start each day with the sun on my face. Ready to hear the birds singing. Ready to leave behind the dreariness of winter. While it has its purpose, winter is not the only season. Thank God. Spring is coming and with it all things new. God always keeps His promises and we can hold on to that, even with chilly, winter hands at the moment. We just need the faith to believe that it won't be long. The sun is going to shine brighter.

Everything's going to be ok. It's a little cold out there today. But not for long. I think I'll set my flip-flops by the door. Just in case. May not be today, but I will need them VERY soon.

"Sow righteousness for yourselves, reap the fruit of unfailing love, and break up your unplowed ground; for it is time to seek the LORD, until he comes and showers his righteousness on you" (Hosea 10:12 NIV).

ENOUGH

Ever get tired of falling short? Not pretty enough. Not young enough. Not old enough. Not smart enough. Not skinny enough. Not curvy enough. Not driven enough. Not rich enough. Not. Not. Not. You could go on and on. The "not enoughs" never end. Haven't we all heard them over and over again? It can be a cruel world sometimes.

Yet the harshest critic can be the one inside your head. There are reasons for that. When you've been told something so many times, you can start to believe it. Your thoughts have the power to control and change you. What you believe is what you become. If you think you can't, you won't. If you think you're not enough, then you won't be. I am tired of hearing "not enough." Sometimes people don't even say it, but I hear it in my head. Sometimes their actions tell me. Then I just tell myself. Sorry, honey. Not enough. Again.

But no more. There is power in your thoughts. Power to make you or break you. Power to reach your dreams or give them up. Power to be all that you can be or just settle for the status quo. Who will you give that power to? I think I will hold on to mine, thank you very much. I don't want to give anyone that much power. I may not measure up to your standards, but that doesn't make me not enough. I am not defined by what anyone else thinks about me. Or I shouldn't be. I am defined by how I define myself. How I think about myself. Don't ever give anyone the power to make you feel "not enough" in any way. We are all imperfect. We are all flawed. We are all human. We all carry around our own "emotional baggage" that really makes us unqualified to judge someone else's "luggage." Who made us to be judges anyway? Your journey is uniquely your own. No one else will ever take the exact one you are taking.

This journey shapes who and what you become. And it makes you...uniquely you. The joy and pain, the sun and rain, the successes and heartbreaks all contribute to who you are. You may not be what someone else wants or thinks you should be. That's fine. I consider that their opinion. Everyone has one. But, that doesn't make it right. Opinions are based on emotions or how someone else "views" something. We all view things and people in our own way...based on our own experiences and desires and feelings. Not always accurately. Not always right. Not always fairly. Not always unbiased. We're *human*. We make mistakes. We misjudge. We don't always see things clearly.

That's why when it comes to determining if you are enough. *you* should be the one that decides. I have let too many people make me feel like I was falling short. But in reality, their words or actions should never have had enough power to override what I know to be true. That I am enough. Maybe not for them. But for me. And for the destiny that is uniquely mine. I am enough. You are enough. Look in the mirror today and tell yourself that. "I am enough." You are a work in progress. We all are. Wherever you are in your journey, remember that you have what it takes to get where you are supposed to be. To become all that you were meant to be. Because you are absolutely—enough.

"And he said unto me, My grace is sufficient for thee: for my strength is made perfect in weakness. Most gladly therefore will I rather glory in my infirmities, that the power of Christ may rest upon me" (II Corinthians 12:9 KJV).

"For God hath not given us the spirit of fear; but of power, and of love, and of a sound mind" (II Timothy 1:7 KJV).

CRACKED POTS

Remember that episode of the Brady Bunch where Peter breaks Carol's favorite vase? He was just told not to play around with the ball in the house, too. The story revolves around trying to get the vase put back together without being discovered. Of course, in the end, the vase sprouts a leak...well several leaks, and the truth comes "spilling" out.

You may think I am going to tell you about the importance of following rules, listening to your mom, or not trying to "hide your sins" because "your sins will find you out." All of those are important truths in themselves. You would do well to remember and respect them, because they can save you a lot of heartache on this road of life.

I really want to talk about being a cracked pot. You didn't see that one coming, did you? You wonder where this is going now. Everything in my life, big and small, has become object lessons from God to me. Merriam Webster says that an object lesson is "an example from real life that serves as a practical example of a principle or abstract idea." It can be the smallest, most mundane thing and He will somehow bring out a deeper meaning that makes me open my eyes to view things in a whole different light.

Using the Brady Bunch may be pushing the definition of "real life," but hey, He made a donkey talk. Enough said.

Let's take this from the perspective that we are all pots. Work with me here. Imagine your "pot" self. As we go through life, our experiences affect our "pot." There are good times when it's like someone is polishing you. You get compliments that make you "shine." You get a new job. You buy that new house or new car. You meet someone special. Get that "A" on that paper you slaved over. (I can relate to that one.) You are up there on that shelf of life sparkling and gleaming in the limelight. You want to shout out "Look at me. I'm the king of the—pots?" I think you get the picture.

Then there are times in this life when you're not treated with the respect you deserve. That unkind remark may chip a little of your paint. That person you thought was a friend betrays you. A little crack appears in your porcelain. You have to struggle to make ends meet. It's hard holding things together. Next thing you know... your handle falls off. What good is a pot with no handle?

That person you thought was special enough that you took the chance of putting yourself in their "hands," ends up dropping you. You really didn't measure up to their standards, so you find yourself on the cold, hard floor of reality. Crash! Suddenly there is not one crack—there are *lots* of cracks. You are afraid to move too much. You might fall apart. After what you have been through, there isn't much holding you together.

Now you get the picture. The cracked pot. What hope or use is there for a cracked pot, especially one that has seen more than its fair share of misuse and neglect? If people in this world aren't happy with how they look, they pay surgeons to make them over, or improve them, so they feel good about themselves.

I guess you could do that with pots too. Slap enough glue and putty and paint on and you may be able to cover the scars that life has left on the pot. Because, what good is a cracked pot? I think I may know. You see, I'm a cracked pot. Really cracked.

Life at times has been hard, sometimes cruel. It has left me with more than my fair share of openings, gaps, and crevices. It could make you bitter if you let it. It can make you sad. Life just hurts sometimes. But I have found that there is a use for cracked pots. We may not be what we once were, but we can be something else.

If you put a candle in a perfect pot, what would happen? Nothing. There is no way for the light to get out. But put one in a cracked pot—there's a different story. The more cracked the better. Every little space allows the light to shine through. A cracked pot can show more light than a perfect pot, because it allows more light and hope to shine through. Someone that has been broken and found the hope to keep going can light the way for others that are looking for a reason not to quit.

Everyone wants to be perfect. But perfection is overrated. It's a high pedestal to fall from, and dangerous, especially if you're a pot. I may be cracked, but I am finding there is a beauty in that. In a dark world, I can add a little light of hope. That, my friend, is more beautiful than being perfect.

Put the glue away, Peter. I don't mind the cracks after all.

"When his lamp shone on my head and by his light I walked through darkness!" (Job 29:3 NIV).

"Therefore, if your whole body is full of light, and no part of it dark, it will be just as full of light as when a lamp shines its light on you" (Luke 11:36 NIV).

"Neither do people light a lamp and put it under a basket. Instead, they set it on a lampstand, and it gives light to everyone in the house" (Matthew 5:15 Berean Study Bible).

The Prodigal

Pigs are cute, especially the little ones—all pink and chubby and squealing, with those little snouts and floppy ears. What's not to love about a pig? Thoughts of pigs have been on my mind lately. I have been thinking a lot about the prodigal son. Prodigal is not a word you hear too often these days, but most of us know the story.

If there was ever a story you can easily look at and become a judge, it's the parable of the prodigal son. Admit it. Just the name has a negative vibe. Prodigal. Admit it. He had it made. A wealthy family. Baby of the family. Probably a little spoiled. No real cares from the sounds of things. I have read and heard the story many times, and have always thought he was so selfish, and immature, and stupid. See? It's so easy to pass judgment. Yet lately I have looked at the story a different way. It's easy to change your point of view when you find yourself in someone ease's shoes—or sandals as the case may be.

Have you ever been at a place in your life when you needed a change? Something different from the life you had always known? I'm not talking midlife crisis here, because the prodigal son was still young. Maybe he wanted to see what was out there. Maybe he wanted to start a new life on his own. Maybe he just wanted a change. For arguments sake, let's say he wasn't selfish and impetuous. Maybe he just wanted to grow up. He wanted to be on his own. He wanted to make his own decisions. He wanted to see what that world out there was like. I can relate.

When you've never been away from home, you wonder what it's like beyond what confines you. You've heard all the stories. It's a big world full of fun and adventures where you get to choose your own destiny. Make your mark. Get your piece of the pie. You want a chance to experience that. So, I can understand why he wanted to leave. When you have heard of freedom, but never really known, you long for a chance to experience it. I don't believe that the prodigal son's desire to leave was a sudden decision either. It says he asked for his portion, and not long after, packed up and left. Not immediately though. There was a little while before he left. But he did leave.

I bet he was excited. It is when you start out on your own headed into a world full of opportunities just down the road. I know that feeling. For the first time in your life, you are in charge. Of you. Of everything.

It's empowering to someone that's never felt they had any power. And a little overwhelming. And a little scary.

But it's a good scary. Because for once in your life, you are the boss. So he heads down that long road. He may not have even set out to be riotous and rowdy and wild. He might have just wanted to have a little fun without the strictures that had always been over his head and life. Maybe he just took up with the wrong crowd. When you have been sheltered from the realities of life, you can be naïve. You just think all people have good intentions and are honest. I see everyone like that. Good. Because of that, I have been handed many cruel lessons. But I refuse to let it jade me. Yes, there are people that have taken advantage of my trusting nature. I have always been too trusting. Believe the best of everyone. Even after I have had my heart walked on and handed back to me, I still believed.

Maybe the prodigal did too. Maybe he thought they were really his friends. Maybe he didn't realize they were taking him for a ride. They were using him. They were exploiting him. When he had served his purpose for them, they evidently left him. High and dry. Or low and broke. Either way he was left on his own. At the same time a famine hits. There go the job opportunities. He did find someone to take him in, and they gave him a job. Feeding pigs.

I know what it's like to look around and find you at the pig sty. It's not a pleasant place. Mucky. Muddy. Smelly. Noisy. Those cute little piglets have grown up to be big and mean. They fight, and they're loud. They push and don't like to share. It's a hard place to find something to sustain you when there are pigs around. They're pigs after all. They eat everything they find. And they'll take what you have, and not care. He was at his weakest point at the pigpen. I've been there myself. No strength. No help. No hope. Desperation.

When you hit the ground hard, do you know what happens? You wake up. You realize it's only the end if you give up there. In the stench and the pig sty. Or you can get out of the mud pile of mistakes. Brush off the stains of regrets. Shake away the sadness. Wipe away the memory of every failure. You don't have to stay in the pigpen forever. You can leave whenever you want. Cohorting with pigs doesn't make you one. It just leaves you dirty and probably a little smelly. But, if you want, you can leave the pigpen behind. It's not too late. You just have to pick yourself up and start home.

That's the great thing about God's mercy. It doesn't matter where you are. Mercy can find you. It doesn't matter what condition you are in. Mercy can change you. It doesn't matter how weak or broken you are. Mercy will hold you. Mercy will wash you clean once again. You don't have to stay in the muck and mire that life can drag you through. Come home. Someone is waiting and watching for you.

"It is of the LORD's mercies that we are not consumed, because his compassions fail not. They are new every morning: great is thy faithfulness" (Lamentations 3:22-23 KJV).

I Have That Shirt

Been there. Done that. Have the t-shirt to prove it. No matter where you have gone or what you have seen or what you are feeling, there is a t-shirt you can put on to let the rest of the world know about it. We will parade around with everything from rock bands, restaurants, favorite vacation spots, and even goofy sayings blazoned across our chest for everyone to see. We are human billboards and we pay good money for the privilege to be one. Little do we realize that all the things we experience in life give us shirts to wear, too. Sometimes they're great. Sometimes they're not. I have a closet full of them myself. Some I wore proudly. Some made me so happy. Some I hope I don't ever have to put on again. There are some I have rolled up and stuffed in the back of a drawer, hoping no one will ever see them. Yeah, we all have those life t-shirts. Life is no respecter of persons. It hands them out whether you want them or not. Sometimes you have no choice in the one you're given. Sometimes your actions earn you one of honor or shame. If there is one thing in this life you can be sure of, it is you will be handed a shirt to wear every day of your life. None of us can escape that. But there is good news. You don't have to accept it. No. You. Don't. You are not defined by what anyone else says or thinks about you. You are defined by who He says you are. You are more than your past mistakes. You are forgiven. You are favored. You are an overcomer. You are loved. You are worthy. You are beautiful. You are not a mistake. You are chosen for great things. He has a plan for you. You are held in His hands. He sees you. He knows you. He loves you. He will never leave you or forsake you. He will make a way. It doesn't matter what t-shirt you've been given. Have joy in the journey. Smile. Trust. Believe. Hope. Not because everything is great. But because He has got you covered. I promise. You don't need a t-shirt to prove it!

"Therefore put on the full armor of God, so that when the day of evil comes, you may be able to stand your ground, and after you have done everything, to stand" (Ephesians 6:13 NIV).

Not Enough

What makes you beautiful? Can you find it in a bottle or compact or a boob job? My friend and I were discussing just this thought when we wandered into the make-up department the other night. I bet anyone that happened to pass us got a laugh if they heard us discussing the products. There were rewinds, concealers, blushes, perfecters, and even camouflages. Need I go on? Sounds like a major undercover operation. And in truth, that is exactly what it is.

But why? Because that is the world we live in. The one that has been trained to look for the beauty in the superficial. The one that has raised a generation of men that feel exactly that way. The one that makes a man believe is he better because he has a beautiful woman hanging on his arm. The one that makes girls, young and old, feel not good enough, because they don't look like the girl on the magazine cover. It's a tough world, unless you look like the girl on the magazine cover.

If you're not beautiful, you can always find something to make you more beautiful. Better make-up. Surgery. Hey, I am not knocking make-up. Kudos to all the women that take the time to make this world a more beautiful place. It's not a secret or a lie. At times we just want to feel prettier.

But my point is why can't men just see the natural beauty that doesn't come from a bottle? Kindness, compassion, gentleness, and sweetness. Those are things you can't buy at a make-up counter. The bald, bandana wearing woman that has lost her hair, but not her strength, to cancer, she is beautiful. The woman that has lost her breasts, but not her fight, she is beautiful. No, she may not have those magnificent tatas men adore, but she has more. She has life.

Natural beauty comes from within. No matter how beautiful you can make the outside, this body is not really who we are. It only houses that person. It must be nice to live in a beautiful "house." I know I will never make the cover of Glamour magazine. I marked that off my bucket list long ago. I just want people in my life that can see past the imperfections and "not enoughs" to see a person that loves without limits, one that gives without questions, and one that cares too deeply. You never have to doubt where you stand with me.

I know there are real men out there that still appreciate that. Superficial beauty doesn't last forever. True beauty does. I want my beauty to come from who I am, not what I have to put on to be enough. Because I am more than enough. You are more than enough. We were designed to fill a purpose that no one else can. Don't ever let anyone make you feel less. My days of not enough are over. If you can't see who and what I am worth, then it's your loss. Because, I am more than enough. You are, too. Don't let the world tell you otherwise.

"Charm is deceptive, and beauty does not last; but a woman who fears the LORD will be greatly praised" (Proverbs 31:30 NLT).

It's All In The Wanting

Maybe it's this time of year, but I have found myself facing some hard truths, and discovered some things about myself and human nature I didn't know. Imagine that. There is always something new to learn. I have been the small town girl in the big city, so to speak, for a while now. It wasn't just the newest changes in my life that have made me see I still have a lot to learn, but's also the gradual ones over the past few years. We all have to find our way out there in the big world.

The wee ones I have spent so much time around have taught me so much about life without even knowing it. Very early in life, we learn the desire of wanting. You want the ball your friend has. You want the candy at the checkout aisle. You want the toy you put down twenty minutes ago, but now that someone else has it, you realize you want it back. The want. These episodes can be followed by grudging acceptance, angry outburst, tug of wars, or full-fledged fits. These are the times when someone older and wiser steps in and helps you realize that we can't always have what we want when we want it. It's not like you really *need* it. Once that statement registers in the young mind, they realize the power of need.

A need is a state of necessity, requiring supply or relief. Now it's a whole new story. Needs require action and fulfillment. Needs must be met. Some needs are necessary just to survive. Then the little mind's realization comes that if they say they need something, not just want it, the importance factor tips a little more in their favor. "But I *need* it. I really do. I can't live without it!" Sometimes it works. Sometimes it doesn't. In fact, once this strategy is learned, we never fully remove it from our repertoire of bargaining skills. No matter how old we get, that little child standing with a handful of candy bars at the checkout, pleading as tears roll down our face, lives on in our hearts and sometimes our will. But I have come to find out that there is a big difference in wanting something... and needing something.

Were you ever one of those middle school kids standing or sitting at the dances, hoping and praying your secret crush would choose you? Giggling with your friends. Waiting and secretly watching his every move anytime another slow song would start hoping that he would head your way and ask you to dance? Oh, I am having flashbacks remembering how that felt. Torture and ecstasy all rolled into one night. He didn't need to dance with you. He would survive without it. But you weren't sure you would. You just hoped with all your heart he *wanted* to.

Remember in gym class when the time came to pick teams? Who hasn't stood secretly praying not to be last? Been there. Done that. More than once. The team may not need my skills, but please let the captain *want* me.

There is a difference in being needed and being wanted, too. But you never really know what you really need and what you really want until you are pushed and broken beyond recognition. When all that is left is tattered pieces and scattered dreams, it is at that point you have to decide to pick up what is left and start new. That is a scary and hard place. I have walked through it. Some places I have crawled through. I have put the scraps I could salvage in some sort of order. I tried to find a peace in the chaos. In those depths of despair, I found out who I was and what I was. What I need and what I don't need. What I want and don't want.

At one time I thought I needed so many things in my life. There are the necessities we need to survive. They're not a choice. They are part of life. But now I see that it's the "wants" that are even more powerful than the "needs." The little ones had it right in the beginning. The desire of a want is much deeper than the desire of a need.

I'm not going to worry about what I need. That will take care of itself. I will consider carefully what I want, because that's where hearts and dreams and lives are made or broken. The God that made the universe doesn't need me. No, I am pretty sure there are plenty of people with much more to offer than I have. In all honesty, He doesn't need any of us. But you know what I realized. He wants me. Yes, me with my failures, my insecurities, my imperfections. He wants me. That's powerful.

He could choose anyone. But He chose me. And you. He didn't leave you as a wallflower at the dance or last pick. No, He chose you. He sees your worth. He knows your value. Even at times when you can't see it. He does. When you think no one wants you, He does. Some may forget you, but He never will. He remembers you every time He sees the scars on His hands. Marked forever because He wanted you that badly. No one else has ever or will ever want me that badly.

"The LORD appeared to us in the past, saying: "I have loved you with an everlasting love; I have drawn you with unfailing kindness" (Jeremiah 31:3 NIV).

"Can a woman forget her sucking child, that she should not have compassion on the son of her womb? yea, they may forget, yet will I not forget thee. Behold, I have graven thee upon the palms of my hands…" (Isaiah 49:15-16 KJV).

Words That I Hate For $500...Waiting

I hate sitting in waiting rooms. I'm not a person that does well sitting still. I need to be busy. Do you know what the worst thing about waiting is? You never know how long you will have to wait and that is the killer. I have spent more time waiting than time spent in the actual appointment.

Sometimes we go through periods in our lives where we seem to be waiting. We are at a place where we are headed for an "appointment" so to speak, but are not quite there yet. You are on the sign-in sheet, but your name hasn't been called. So you have to wait. And oh, I hate to wait. It's not that I'm impatient. It's like I said before. I just need to be busy. Waiting can be more tiring than being busy. But sometimes we are stuck waiting.

At least that used to be my mindset. Then God gave me a different insight. I told God this seems to be taking so long. He gave me promises. I know He will keep them, but it seems like the wait is just taking too long. Then He said, "Who said you were waiting?" Now what does that even mean? Of course I'm waiting. Then He said, "No, you are not. When you wait, you are still. You have not been still. Every day you have been moving closer to what I have for you, to becoming who I want you to be. You have never been in the waiting room." Wow! It's true, so true. I realized it has never been about waiting. It has been about preparing. We can't see what God has for us, but He knows what He has to take us through, and what He has to take out of us to get us ready for what is ahead. He is getting rid of everything that shouldn't be, and opening the doors to what should.

So we're not waiting, we're preparing, and that changes the whole way this thing looks. So don't be discouraged.

I believe it will all make sense soon. It will make perfect sense because He can only do perfect. Until it all comes together keep going. It is going to be worth it. I just know it. He promised:

"Be still, and know that I am God" (Psalms 46:10a KJV).

"But they that wait upon the LORD shall renew their strength; they shall mount up with wings as eagles; they shall run, and not be weary; and they shall walk, and not faint" (Isaiah 40:31 KJV).

Finding God's Heart

I have always wanted to be a woman after God's own heart. I think that's why I have always loved the stories of David. We think of him as the shepherd boy, the giant slayer, the mood singer (he could drive out bad spirits with his songs) the best friend, the king. But sometimes we forget he was a fugitive, an outcast, an adulterer, a murderer. Yet the Bible calls him a man after God's heart.

How can that be? What is there about David that separated him from so many others whose stories we have heard? He was a man after God's own heart.

David was well acquainted with his human side. One of the things I have learned from David is that no matter what the sin or how far the fall, he got back up and tried again. There were many times in David's life he could have said "Forget it. I'm done. Why bother?" But he didn't. He knew his weaknesses. At times he even gave in to them. But he knew something else. He knew he needed God. Like a deer needs water. I need Him. You need Him.

He knew God saw everything about him. His ups and downs. His successes and failures. His good and his bad. I know he sees each of us at our best and at our worst. We are so far from perfect. We battle this flesh daily. Sometimes we win the fights. Sometimes we lose. But just like David, no matter how many times we fall, we need to get back up. No matter how many times we fail, we need to keep trying. No matter what others say or think about me in the big scheme of things won't really matter. Just let this one thing be said. Let it be said I was a woman after God's own heart, and please God—let it be true.

"As a deer longs for flowing streams, so my soul longs for you, O God" (Psalm 42:1 GWT).

"Whom have I in heaven but you? And earth has nothing I desire besides you" (Psalm 73:25 NIV).

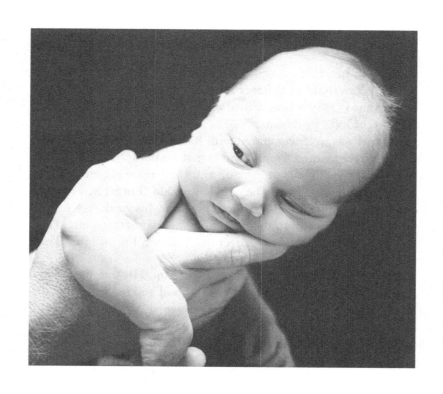

You'll Get Through This

You'll get through this. You will. I know it's hard. I've been there. Sometimes heartbreak and disappointment can be so devastating that it actually physically hurts. I honestly believe a heart can break. Mine has. The pain was so bad I could barely breathe. I had to tell myself to keep breathing. Every single breath took effort. I prayed to God for strength to just keep breathing. And

He gave me that and more. I didn't think I could make it past the pain. But I did. You will, too.

I learned a lot about breathing from being in pain. Breathing serves an important purpose in life. It lets us know we're still alive. Pain does the same thing. You'll find that breathing and pain can actually work together. When you learn to control your breathing, you can connect your mind and body. No, not turning yoga instructor—yet. But, there is a lot to be said for the benefits of learning to breathe. If your mind can take over and decide where to put your focus, the body can handle the pain. (Just thought I would add that it works for stress and tension, too. Maybe the Yoga instructor *is* surfacing.)

The simplest way to explain this is by looking at breathing in the childbirth process. It's not your mama's hee hee hoos either, although they do serve a purpose. When the labor pain starts, it's fresh and sudden and sharp. Like heartbreak or disappointment or loss. Meet the cleansing breath. Or the signal breath. It lets you and everyone else know you are having pain and are ready to focus—away from the pain. You take a deep inhale through your nose, holding through the pain, then release through your mouth. In the beginning you are just trying to relax and come to terms with this hurting deep inside. It is probably like nothing you ever felt before. It is a ripping, searing pain.

There are things in life we face that cause that unimaginable pain. I have felt it. Maybe you have to. I got through it. You will, too.

Because next you have to face active labor, which is, exactly what it sounds like. Hard work. Enter the hee hee hoos. Sometimes concentrating on breathing isn't enough. Painkillers enter the picture. They dull the sensation of pain. Some people have trouble getting through the pain…they need something to dull or kill the pain. They grab onto anything or anyone to take their mind off the hurting. They want to ease the ache and forget, even for just a little while how much they are suffering. But, the pain is still there. It is just hidden or masked by the effects. If you can allow the pain to do its job, you will be that much closer to reaching the place where the pain is over. So you hee hee hoo with everything you have. You even hoo hoo hee if necessary. You keep breathing. Faster. Harder. You do whatever it takes to take your mind off the pain.

You reach the final stage. You no longer fight the pain, but welcome it because it will bring the release you need. The end is near. No holding back or avoiding it now. No. The more you accept it and let it work, the faster you can move out of the pain. When you get through the pain—there will be joy. Just like when that baby finally arrives. The excruciating pain you experienced is forgotten. You made it through. There may be scars and memories of the pain, but you survived. The process will change you. Permanently. Pain will do that. It makes you realize your weaknesses. And strengths. It shows you where to place your boundaries. And where to let them down. It teaches you who you can trust. And can't. Make no mistake. It hurts. But it serves a purpose. Pain is an excellent teacher.

If you learn well, you won't have to repeat the same painful lesson twice. If you don't, you will. Just a fact of life and its many lessons. Just remember. Whatever you are facing, you'll get through this. You will. Just remember to breathe.

"A woman giving birth to a child has pain because her time has come; but when her baby is born she forgets the anguish because of her joy that a child is born into the world" (John 16:21 NIV).

"For our present troubles are small and won't last very long. Yet they produce for us a glory that vastly outweighs them and will last forever!" (II Corinthians 4:17 NLT).

Cowboy Up

A trip to the rodeo can prove to be an interesting adventure. Being a horse lover, I enjoy watching all the different horses prancing and pawing and performing. It makes me want to get out there in the saddle and stir up some dust myself. Another exciting event you will find at the rodeo is the bull riding. Most bulls in these events weigh an average of 1,500 pounds. That's a lot of bull. Some guys actually ride bulls for a living. It sounds easy. Stay in the saddle eight seconds and you are successful. We have watched some rides where bull riders hung on for dear life. Other times they knew it was time to "cut rope and run," or at least let go of the rope and jump out of harm and the bull's way. I guess it can be a thrilling and dangerous job. We watched rider after rider getting tossed and trampled. That could be discouraging if it happened often. It would be easy to become disenchanted with life after having to pick yourself up over and over again. But that's the life of a bull rider. Hopefully, they don't become jaded from the falls.

In a world that embraces the cynical, I refuse to become jaded like that unsuccessful bull rider. It is easy to let the struggles and hard places we face leave us feeling bitter and dusty as a tossed rider. Life can be unfair at times. It can be heartbreaking. It can hurt. But it doesn't have to change how we look at life or ourselves.

We can't control the things that come our way, but we have the choice of choosing how we respond to them. It may not change the situation, but it will change us. When you change your mindset, you can change everything else. A change in how you look at things will give you a new strength to rise above and find peace in knowing that no one can keep you down—except you. Don't give anyone or anything else the power to control you or what you believe.

You have the power to choose. I know what it's like to be hurt and broken. Haven't we all been in those places? But even after the pain, we can still choose to hope and believe that the story is not over. The chapter of that part of your story may be, but do you know what follows the end of a chapter? The beginning of a new one. Our lives go on. We don't have to let the past define us, but it can refine us.

Like the bull-riding cowboy, it's all about how well you

stayed in the saddle or let go when you needed to. If you can get back up, you're doing good. You may be a little dusty and bruised, but that is a temporary state. You made it through to ride another day.

The things we go through in the past help mold us for the better. Stronger. And isn't that the way it should be? So instead of letting things get you down, choose to look up. Choose to believe. Choose to see the good. There is always some if you look for it. Choose to be happy. Choose to hope. Choose wisely. It's all about how you choose to look at things anyway. In the end you will see that it wasn't about the hand that life dealt you, but how you played it. Or rode it. Cowboy up.

"You will keep in perfect peace those whose minds are steadfast, because they trust in you" (Isaiah 26:3) NIV.

The Miracle Of Making It Through

We all have circumstances to face in life. Sometimes God steps in and performs a miracle of deliverance—lifting us out of the fiery trial of our faith. We see how He can make a way where there seems to be no way.

At other times He performs a different kind of miracle. Not by removing the obstacles we face, but by giving us the strength to keep going. In the face of adversity, we learn to trust Him. We learn to hold onto His guiding hand—even though we may not understand where He is leading us.

We learn to believe, not because things are good, but because we are standing on the promises of the One who cannot fail. We learn to trust that He who began a good work in us is able to keep us until the work is done. We learn to keep walking faithfully. When learn that when we are going through storms, but can still reach out and help someone else that is struggling, a new strength comes.

It is here that we find a different kind of miracle. The miracle of making it through. When we make it through, there is new strength. There is new victory. New purpose. We can thank God for every step that leads us closer to Him and the plan He has for our lives.

If you are in the middle of the storm or the trial, hold on. He never takes us where He can't or won't keep us. It may be hard right now to see how this will all work out. Just keep going through. One step at a time. One day at a time. You will find your way. He promised. I can't wait to see what is going to happen *when* we make it through.

"So be truly glad. There is wonderful joy ahead, even though you must endure many trials for a little while. These trials will show that your faith is genuine. It is being tested as fire tests and purifies gold—though your faith is far more precious than mere gold. So when your faith remains strong through many trials, it will bring you much praise and glory and honor on the day when Jesus Christ is revealed to the whole world" (I Peter 1:6-7 NLT).

"Dear friends, don't be surprised at the fiery trials you are going through, as if something strange were happening to you. Instead, be very glad—for these trials make you partners with Christ in his suffering, so that you will have the wonderful joy of seeing his glory when it is revealed to all the world" (1 Peter 4:12-13 NLT).

No Good Deed

I truly believe that no good deed goes unpunished. Now you may think you know where I'm headed with this one. Wrong. Instead of making you look over your shoulder in fear of what may be retaliating, I want you to look ahead with hope. I have tried to live my life transparent. Even long before someone else suggested transparency in politics, I have always tried to be what I profess to be. What you see is what you get. Most of the time I have succeeded. Not always. I am undoubtedly human. There have been times I have failed miserably. But still I try.

My daughter always told me my kindness would get me nowhere. Still I have been kind. I have been told I am too nice. Still I am nice. I wear my heart on my sleeve. You don't have to wonder where you stand with me. There have been times people have taken advantage of that. That really hurt. It hasn't stopped me from living that way. Oscar Wilde said "Life is not complex. We are complex. Life is simple, and the simple thing is the right thing." No game playing. Honesty. Being who we say we are. Simple right? We make life so complicated when it doesn't have to be. We try to live up to standards that are so high we struggle to hold on to them without losing our grip and falling. Or we don't even try to be what we could be. I have been on both ends of that spectrum. We all have, if the truth be told.

Wouldn't it be great if we were all the person we put out there for the world to see? You know the one you want all your Facebook friends to think you are? The cool, popular person that has it all together. The one with the great profile pictures and the funny or timely posts that your friends all "like." But in reality, most of us fall somewhere in between. We need to be true to ourselves. Think how easy life would be if we could all be ourselves. But, we can't be. Or we think we can't.

People will judge us. People will use us. People will hurt us. Guess what? They already do. "To thine own self be true..." Oh, Shakespeare, you had it right. The most important person we can be true to is our self. God sees everything we say and do and are anyway. He already knows exactly who we are, no matter what persona or costume we wear when we face the world. He knows the thoughts and intents of our heart, too. The Bible says we reap what we sow. Said another way, what goes around comes around. It's that action or deed that produces a reaction.

Long before karma became popular, the Bible told us "Don't be deceived. God cannot be mocked. A man reaps what he sows" (Galatians 6:7 NIV). The things we say and do are like throwing boomerangs. They eventually come back at you—sometimes at full speed. If you are always kind, does that mean you will always be treated that way? Unfortunately no. But be kind anyway. Because it does balance out.

The other day I experienced the "sower/reaper rule" for myself. I had a need. A big one. God had told me to do something. I wanted to be obedient, but there was no way I could make it happen on my own. I prayed that God would supply. I just couldn't see how. But I believed. I stepped out in blind faith.

I got a message on Facebook from a friend I hadn't seen in a while. She heard I was moving and wanted to visit before I left. I had cared for her son during his preschool years. We met to catch up, and reminisce. As they left, she handed me a card. With the holidays near, I didn't think much of it. After they left, I opened the card. It said, "Thank you so much for all you have done for us and being such a constant source of inspiration." And there was a check. God supplied. They didn't know my need. But they had listened to God. I prayed and God answered. I hadn't done anything special in my eyes. I had just tried to be kind and giving every day. There were many humbling tears that fell. When I talked to her later and explained the miracle, she said prayer is the original and best social working network.

How true. I say none of this to make myself seem better in any way. I have made mistakes. Big ones. There's that human factor again. I just can't seem to get the equation right every time. But I want you to understand that when we give our best, do our best, be our best, it all balances out. Some people may not care. Some people may not understand. Some people may even misuse what you offer. Do it anyway. The right ones won't.

Instead of no good deed goes unpunished, maybe no good deed goes unrewarded would be a more fitting ending. The One that keeps the balance sheet in life knows and sees. His word doesn't lie. You reap what you sow. If you plant roses, you will get to roses. If you plant kindness, kindness will blossom in your life. Choose your seeds carefully, but plant them everywhere. You will be happy when you see what grows.

THE ABYSS

Lately, I feel like I am standing on the edge of a mountaintop, the winds battering me from all sides. The force is so strong, I have trouble holding my balance. It looks like a long way down. I can't really see past where I stand—just a great abyss. But I can see something is coming—the storm. It looks like a bad one.

    I hear a voice calling "Come." I hesitate. I don't know what's out there. Is there anything out there? What will I be stepping into? The fear of the unknown can be paralyzing.

Turning back is not an option. Neither is stopping. But where am I going from here? There is nothing to calm my fears. It looks like I am headed into the storm.

They say an eagle knows when a storm is coming. Long before it breaks, the eagle will fly to a high spot and wait for the winds. When the storm hits, it sets its wings so that the wind will pick it up and lift it above the storm. The storm may be raging, but the eagle is soaring above it. The eagle doesn't run from the storm. It simply uses the storm to lift it higher. It rises in the winds that bring the storm. I must learn to do that, too.

He speaks to me again. "Come." Storms can be scary when you are out in them with no protection or shelter. It looks like that is where I am. Alone. Unprotected. Powerless. Vulnerable. Helpless. Unsafe. Exposed.

But, that isn't really true. In this life we never have to be alone. We never have to feel there is no safe place to run. "God is our refuge and strength, a very present help in trouble." Psalm 46:1

I know He won't let this storm overtake me if I can trust Him. He has never failed me before. He won't now. I can't make it without Him. That is the simple truth. It may look scary ahead, but it would look even scarier alone without Him. It is here where I must trust Him. Again. When you step into the unknown, one of two things will happen. He will catch you, or you will learn to fly.

I have to make the decision. We all have a free will, and He will never overstep that boundary. I have to step out in faith, knowing that He will lift me above the storm, and I will soar to new heights. After all, it's the release that will give me wings.

"Come" He whispers. And this time, I do.

"When the storm has swept by, the wicked are gone, but the righteous stand firm forever" (Proverbs 10:25 NIV).

"Trust in the LORD with all thine heart; and lean not unto thine own understanding. In all thy ways acknowledge him, and he shall direct thy paths" (Proverbs 3:5-6 KJV).

WHERE ARE MY GLASSES?

I might need glasses. I am reading more than ever now—if that's even possible. I have always been a reader, but now I have *big* textbooks with *big* and *long* words. I spend a lot of time with my face in a book. Things look a little fuzzy after a while. When I look at my life, there are times when things look fuzzy there, too. What was I thinking when I decided to take that path? Whatever possessed me to do that? Say that? Buy that? Wear that? Yeah, there have been some questionable choices made on more than one occasion in my life. My vision seems a little impaired. Oh, the weight of being human.

But the other day, it was like I suddenly had spiritual glasses and could see things in my life in a different way. We are like the little bird that stops flying and lands on a post to take a closer look at something. We get a new perspective because things always look different when we change the viewing angle. I thought of situations that could've crushed me. People that meant to hurt me. Roads where I took a wrong turn. But as if a light had been turned on in a dark room—it all became clear. I could see clearly. So many things made sense. I understood how God had taken the darkness, the sadness, the desperation and turned it into hope, joy, and a new purpose. One of my favorite scriptures is "All things work together for good to those who love God, to those who are the called according to His purpose" (Romans 8:28). I may not always do the right thing, say the right thing, and choose the right thing. But we have the promise that He will work all things out for our good. Because we love Him. When we are where we should be, His goodness and mercy will follow us. When we are trying our best, but fall short, He will turn it around for our good. Even when we make mistakes and fail, He will turn it around for good. What a promise! And there are those that for whatever reason, meant evil against me, but I have seen that God can change the bad to good, and He turn the situation.

Don't be surprised if you see me in my pretty pink "spectacles" now and then. I still have a lot of reading to do, and these human eyes need a little help seeing clearly once in a while. I hope when you look at your life, you'll see past the fuzziness of appearances, and get a clear view of all that He is doing. Turning it around for your good. Seeing is believing. You just have to open your eyes.

"God will make this happen, for He who calls you is faithful" (I Thessalonians 5:24 NLT).

"Open my eyes that I may see wonderful things in your law" (Psalm 119:18 NIV).

TANTRUM LAND

I have spent many days with preschoolers. When you work with "little people," you become well acquainted with temper tantrums. I can honestly say I have seen every type of tantrum known to the human race and lived to tell about it. Some can be deterred with simple reasoning. Some can be stopped with a stern "That's enough. We are *not* doing that!" And yes, there are some that you simply have to ignore. You step over the screaming, thrashing child on the floor, and calmly go about your business until they exhaust themselves (or you can't take it anymore, whichever comes first). It is then you can pick them up and talk some sense into them. Explain to them the errors of their ways and help them to see how you know best.

I love that God has a sense of humor. Lately He has shown me that I have been having spiritual temper tantrums. I have been channeling a hard-headed, strong-willed preschooler. Things in my life aren't happening in the way I thought they would or in the time I thought it would take. Do I believe God will do what He has promised? Of course. Yet—I don't want to wait. I want it. Now. So I stomp. I drop and start kicking my spiritual feet. I am crying out at the top of my lungs with all my heart. At first, the Lord tried to calm me. I knew He was there, but I ignored Him. Next He spoke to me. "Be still, and know that I am God." That scripture appeared in my Bible reading; in my daily devotions; and even on my teabag tag at a friend's tea party. Seriously. It was the topic of Wednesday night Bible study. I was bent down looking at jars of preserves at a church bazaar. When I stood up, there above the shelf of preserves was a cross stitched sampler for all the world to see. And can you guess what it said? "Be still and know that I am God." You would think I could get the message. But still I inwardly raged on.

Finally it reached a point where the Lord stepped out of the way and let me go at it while He patiently waited. I gave it all I had. I wasn't happy and I made sure I let Him know it. So I raged and raged in Spiritual Temper Tantrum Land until the fight finally went out of me. I was finally quiet. Finally exhausted. Finally surrendered.

Then He picked me up, like a Father, because I am after all, His child. He wiped my tear-stained face. His calming words spoke peace to my weary, battered heart. He knew what to say to me (just like I would say to that fit-throwing two-year-old) to help me accept that even though I may not have understood, He was making all things work out for my good.

We know He loves us. We know we can trust Him. So now we can calmly climb down from His lap and get ready to try again. We can look to Him, before we take off, happy to know that He is watching out for us and over us. We may not see how things will all play out, but our Father has it all in control. It's safe for us to move ahead because He's got our back. (And to be honest, tantrums are exhausting!)

"When I was a child, I talked like a child, I thought like a child, I reasoned like a child. When I became a man, I put the ways of childhood behind me" (I Corinthians 13:11 NIV).

"No discipline is enjoyable while it is happening—it's painful! But afterward there will be a peaceful harvest of right living for those who are trained in this way" (Hebrews 12:11 NLT).

I Have A Friend

I have a new friend named Harry. He has big brown eyes and black hair. When he smiles, his whole face lights up. His laugh makes me laugh, too. He is a whole two years old. And Chinese. When he came to school, the only thing I could understand that he said was "Mama"—while tears streamed down his chubby face. It's not unusual for newbies to cry. They find themselves in an unfamiliar place with unfamiliar people, and in Harry's case—we didn't speak his language. The adjustment period is different for every child, so there can be a whole lot of crying going on sometimes. Working with small children takes compassion at times. I am softhearted anyway, but there was something about this one that touched my heart.

As time went on, even without speaking, I began to understand what was bothering Harry by his cries. The sad. The mad. The frightened. The tired. He didn't need to speak to tell me. I just knew. Sometimes I could help him. Other times I just held his hand and said soothing words. One day as I sat by his cot at naptime—holding his chubby hand and rubbing his head as he fell asleep, I thought of how many times God has been by my side, holding my hand. I thought of times when the pain was too great to even put into words as tears streamed down my cheeks. Times I couldn't even find words to express the hurt I felt. But it didn't matter. He knew my every cry. The heartbroken. The anguished. The desperate. And He understood—even without speaking a word. He just knew what my heart couldn't even say. And He stayed. He held onto me when I needed Him so desperately.

You will be glad to know that my friend, Harry, is happy now. He laughs and plays, and has picked up English quite well. *And* I am happy, too.

Change is hard—no matter what your size. But through change we learn to see that there is so much to hope for. So many possibilities. So many dreams left to fulfill. We just have to take a chance and reach for them. Today, reach out and be somebody's hand-holder. Show them that even when you don't know the words to say, they're not alone. We all have days we need to know that. If you're having one today, come find me...I am an expert hand-holder.

"For I the LORD your God will hold your right hand, saying to you, Fear not; I will help you" (Isaiah 41:13 AKJV).

"If I take the wings of the morning, and dwell in the uttermost parts of the sea; Even there shall thy hand lead me, and thy right hand shall hold me" (Psalms 139:9-10 KJV).

The Test

I failed it again. The test. Did you know that the Great Instructor in the sky requires us to be tested? Oh yeah. And I failed on my latest one—*big time*. I thought I had it nailed this time. See, I realized not too long ago that He gives us tests and if we don't get it right we have to redo the test—again and again—until we pass it. This can be repeated over and over, though not on the same day. No, like a good Teacher, He gives me time to prepare and get ready. I often think I know the right answers. I think I have it figured out now. Then the test comes.

The first part is easy and I'm sailing right through. Of course it should be. *How many times have I faced this test now?* Then I get to some parts that look a little different. Same material, but it's not like it was presented last time. So I waver a little. Lord, why do You have to change it up? Here's where I have the choice. I need to stop and remember what He has taught me since the last time I took this test. (And the many time before it.) Each time I failed, but I came away knowing a little more than I knew the last time.

Yet instead of thinking with my head, I let my heart take over. I have failed this same test how many times now? That little bit of fear over whether I can do it this time, and the next thing I know, I am writing equations that won't work. Reasoning in ways that makes no sense. Guess what? Failing, yet again. I want to leave the class and never come back. I am so ashamed I can't pass this.

They say the teacher is silent during the test. Let me tell you this. They are *not* silent afterward. He shows me in no uncertain terms all my errors. He's not pleased with my performance—again. He tells me that I have been listening. I have been paying attention. But I'm not using what He has given me to move ahead. I keep missing the same ones. He wants me to move ahead—not just to succeed, but to excel. He tells me I am so close, but I can't go until I have learned what I need to know and can apply it. Oh. No. I have to take it again?

Then the Master Teacher tells me a few tricks. "I can do all things through Christ which strengthens me. Be of good cheer I have overcome....I will never leave thee nor forsake thee...God is faithful. He will not allow the temptation to be more than you can stand. When you are tempted, He will show you a way so you can endure." If He believes in me and has given me what I need to work the problems out, then I can do this. So. It looks like I am going to be taking this test *again*. If I want to move on to the next class, I guess I better sharpen my pencil—one more time. I know I can do it this time.

"Examine yourselves to see whether you are in the faith; test yourselves. Do you not realize that Christ Jesus is in you—unless, of course, you fail the test?" (II Corinthians 13:5 NIV).

Believe in The Magic

A reason to keep believing—that's what everyone is looking for. I have always believed in Santa. And the Easter Bunny. And the Tooth Fairy. And unicorns. And dragons. And fairies. And gnomes. And leprechauns. And pots of gold at the end of rainbows. The magic in wishing on stars and the joy of blowing dandelion puffs. I could keep going, but I'll spare you. I am sure none of this surprises you if you know me, or after reading some of my writings.

Magic. In itself, people think of magic as a bad thing. Let's try magical. When you try to define magical it is something so extraordinary that it seems to be it has powers that defy the powers of nature. Hmm. Charmed. Entranced. Wondrous. Remarkable. Doesn't sound bad to me at all. I love the magical. I look for the magical in every day. For the wonder. For the extraordinary in the ordinary.

Most people see so little magic in the world. That makes me sad. Because that's what I look for. Sometimes it's easy to find. Other times it's not. Sometimes I just have to keep believing that it's there. I have even been known on occasion to sit and weed through patches of clover—looking for a four leaf one. True story, and I just did it less than a month ago.

No one ever said life would be easy. Or fair. If they did, they were lying. Big time. It would be more truthful to say that at times life is just unfair. People can be are mean, even cruel. Sometimes without even knowing, and at other times on purpose. Things don't always turn out like you thought or wanted them to. You can face unexpected turns of events, too. Divorce. Losses. Separations from loved ones. Sickness. Death. That's just the reality of life. You never know what you will be facing. You never know how the things you face will turn out.

It's easy to believe when everything is going good. Things are falling into place. The sun is shining. Birds are singing. But it's when the rain is pouring down and your umbrella breaks and blows away in the wind, you have to learn to believe the sun will shine again. It's when there is nothing or no one in your hands to hold on to, you learn to believe that it won't always be this way. It's when you are so sick don't have no desire to get up or the strength to even try, that you learn to hold on to a belief that you will recover. It's when there's nothing in your bank account, you learn to believe that this is only a momentary monetary state, and that things will turn around.

It's when your heart hurts so badly, that you don't think you can go on, you learn to take it—one breath at a time—believing that you can make it past the pain. It's when you reach your lowest point, you learn to look up. After all, where else is there to go from there? You can quit, or get up and keep going. You learn in those dark and difficult times to keep hoping. Keep trusting. Keep believing.

I know what it's like to have nothing left, but hope. But that is all you need to keep believing. You can trust that He knows the plans He has for you. It may seem like chaos now, but God has a way of turning the mess into a message, the test into a testimony, the trial into triumph, and the victim becomes victorious. Even when people mean to harm or hurt you, He can take that and bring good from it. I have seen Him do it. Time after time. So I believe.

Life isn't perfect. Life may disappoint you. People may fail you. People may hurt you. But you can still believe.

Look for the magic. Magic can come in many forms. It's not the sleight of hand or being faster than the human eye can detect. It's not conjuring and spells. It comes in the simple things.

The rising sun lets you know it's a new day with new possibilities and new adventures. There are new chances to start again no matter how you finished the day before. There is magic healing power for a hurting heart that comes from hearing a child's uninhibited laughter. Or hugging a friend. Or speaking the comforting words that soothes a weary and worn soul. It's always within grasp if we look for it.

Make the magic happen. For yourself. For others. Believe—or help someone else believe. You may not find gold coins from the leprechaun's pot, or be sprinkled with pixie dust and gain special abilities, but you can find the awe and wonder in the world around you. Find the magic. Be the magic. I am headed outside now. I saw a promising patch of clover earlier...just by that rainbow's end. I better check it out.

"Now faith is the substance of things hoped for, the evidence of things not seen" (Hebrews 11:1 KJV).

"He staggered not at the promise of God through unbelief; but was strong in faith, giving glory to God; And being fully persuaded that, what he had promised, he was able also to perform" (Romans 4:20-21 KJV).

The Science of Making Toast

I write a lot about things I love, but what about things I hate? Yes, even this Pollyanna thinks there are a few distasteful things in the world. One thing I hate is burnt toast. A toaster is a peculiar gadget. Toasters are almost like fingerprints. You can buy the same brand, model, and make of a toaster, yet you will get a toaster that uniquely gives you a different shade of toast even when you set it at the toasting "number" you always use. Huh? How is that even possible? We can send men into space, but can't get conformity in toasting bread. And for some reason, when I am rushed or busy, I always get the same shade. Burnt brown. Or black. The setting can be where it always is, but the toast ends up a little well done. Explain that.

It seems like things in my life end up like that toast at times. I have really good intentions. I try to get it right. Say it right. Do it right. But somehow things turn. The situation seemed so simple. Like making a piece of toast. No need for a degree in rocket science to figure this out. But somehow, some way, it doesn't turn out right. It gets messed up.

Sometimes the fault lies with you. You were busy. You were in a hurry. You weren't paying attention. You missed the signs around you that things weren't "set" where they should be. You end up with your efforts going up in smoke in front of you. And burnt toast stinks, too. It makes the whole house smell bad. You can leave and return hours later to the lingering aroma of "burntness." It just is hard to get rid of that stink. It clings to your hair and clothes.

Things in your life can do that, too. Leave you with bad odors from the things you've been around. It can be hard to get rid of some scents. Like onions and skunks. You keep washing, but still the smell lingers on. It can be hard to get rid of wrongs and mistakes that are holding you captive like the haze of smoking toast. You try to distance yourself, but you can still smell it. You wonder if other people can, too. You just wanted a piece of toast. You just wanted something simple, but you ended up with a burning mess. Not edible. Not worth sharing. Not worth saving. If it's really bad, you toss it. Once in a while all isn't lost and sometimes you can scrape the darkness off and still use it.

You can do that in life, too. There are times when we make a mess of things. We could "toss it" and just give up or we salvage what we can and make the best of it. Pick up the knife, get rid of what we can scrape off, and use what's left. I have had to do that in life, on more than one occasion. I messed up badly. That meant one of two things. I had to decide to toss or scrape. Most of the time I scraped off what I could get off. It wasn't exactly what I had intended it to be, but I have learned to not waste what I am given. Whether it's the hard lessons in life or burnt toast. It has taught me to be more careful of my choices and watch more carefully what I am doing.

In this life we will never be perfect. Never get it right every time. But that's okay. He has got you covered. I will keep trying, in life and toast making. I hope you keep trying too.

But just in case, it's nice to know there's a knife and forgiveness close by.

"Indeed, we all make many mistakes. For if we could control our tongues, we would be perfect and could also control ourselves in every other way" (James 3:2 NLT).

"Be very careful, then, how you live—not as unwise but as wise" (Ephesians 5:15 NIV).

CAN I HAVE A COOKIE, PLEASE?

I have been thinking about my friends. Friends are truly a blessing from God. I have become so thankful for the people that God has placed in my life. I know there are days I wouldn't have made it without their prayers and help. God makes no mistake when He gives them to you.

Did you ever think about all the different friends you share your life with? It's funny how you can have so many different kinds of people as friends. You know what I'm talking about, right? Some are just like you. Others are very different, but somehow you still get along. Some share similarities in personality and disposition and character. Others are just characters. It's a great mixing bowl or let's say "cookie jar" of friends. And if there is one thing I am an expert on it is cookies. I am a connoisseur of cookies...an invertible Cookie Monster. I love my cookies. Big ones. Little ones. Hard ones. Soft ones. Chewy ones. Crunchy ones. I think you get the picture. Oh no. I am picturing all those cookies in my head.

Since I love and know cookies so well, I think I can explain my friends in terms of cookie varieties. I have friends that are familiar and comfortable like chocolate chip cookies. They are never hard to find and make me feel grounded when the world is rocking. They are the constant in my ever-changing life.

Some are like peanut butter cookies. You can't help craving their company. Nothing smells like peanut butter cookies baking.

The aroma fills the house and you can't wait to get your hands on them. Some friends are like that. You can't wait to spend time with them. No matter what you are doing, even doing nothing is better when they're around. Friends like that are the ones that you look forward to hanging with because they leave you feeling satisfied and happy.

Then there are my "sugar cookie" friends. They are sweet and simple. They are your down-to-earth friends. You don't have to put on a show for them, or impress them. They just accept you as you are—faults and all. They are not complicated or hard to figure out. They are just what they appear to be.

I have some friends that fall under the nutty cookie recipe file. Some of them are crazy. But everyone needs some nuttiness in their life (and in big doses) to help you get through bad days.

I even have a few that are like biscotti. Biscotti is tough and hard to chew. Some of my friends can be like that. They aren't the easiest cookie to eat. But if you know how to soften them up, they can easily become a favorite. Another thing about biscotti friends is that because they are dry, they last longer. You don't have to worry about handling them carefully like some other more delicate cookies. They will stand the durability test. They're tough. They will be around for a long time. They'll have your back.

I even have a few "twisted" cookie friends. They seem to have come from some bizarre recipe, but I find sometimes I need to step out of my comfort zone. The "twisteds" help me do that. They make me laugh at the absurdities of life and myself.

I couldn't pick my favorite friend any more than I could pick a favorite cookie. Some days one kind seems right, and another day I need a variety. Maybe that is why God puts so many different kinds of friends in our lives. No matter what we face, He puts someone in our life to help us through. They all bring different, but necessary ingredients to our lives.

I am so thankful for all my friends. I thank you for the sweetness, richness, and taste you add to my life. For those of you that bring the occasional nuttiness, I thank you for that, too. You all have filled a spot in my heart and life, just as those cookies fill my belly. Now I desperately need to go find some cookies. Seriously.

"A friend loves at all times, and a brother is there for times of trouble" (Proverbs 17:17 ISV).

THE ONLY JESUS

I passed a mirror the other day. "That's the only Jesus some people will ever see." Excuse me? "That's the only Jesus some people will ever see." When you really think about it, it's kind of an eye opening revelation. There are people that will never in their life walk through the door of a church. They will never open a Bible. As shocking as it is, there are people that have never owned a

Bible. Some people will never watch a TV church show. They won't even stop on the channel. You are the only Jesus they will ever see.

There are people around us every day. At school. In your neighborhood. At work. Even in your home. They are watching the life you are living. We all are watching everyone. We know this by the way everyone look at Facebook and Twitter. Even if we aren't around each other every day, because of social media we can be connected constantly. At the touch of a finger. We all are intertwined in each other's lives. We are all watching. But what are we showing them?

We meet people every day that are looking for someone to show them kindness, a little love, a little hope. We have that hope. We know that Hope. We can share that hope with them. I don't mean preaching to them. In fact the life you live speaks louder than any words ever could. You can say you are Christian, but your actions will show the truth. Look at a tiger. You can tell one by its stripes. There is no mistaking it for a monkey. A tiger acts like a tiger. You won't find one swinging through the trees or peeling a banana. They look like a tiger. They act like a tiger.

Remember the old saying do as I say and not as I do? The world is looking for people that are who they say they are. They are tired of fakeness. I'm tired of fakeness. I want to be the real deal. That's not to say that we are perfect. Even when you are trying your best to live for God, you will make mistakes. I fight this flesh every day. Sometimes I win, sometimes I lose. But that doesn't means it's over. It just means I'm human, and I am going to try to do better tomorrow. That is what people are looking for. Not perfect Christians. They are looking for real Christians. People who are like them. People that still struggle. People that know what it is be a sinner and know that they can be forgiven. We can show them His love. We can show them that His blood will still wash away every sin. We can show them that no matter how dark it is in their life—He is light. We can show them that they're not alone. He will never leave them or forsake them. It doesn't matter how weak they are or how they struggle in their weaknesses. He will be their strength. They don't need someone pointing out their faults and judging. That's not Jesus. Don't get me wrong. God hates sin because He knows what sin does to us, but He loves the sinner. I want to be a saint. Some days I am just a sinner. We all have sinned, but what they need to know is that someone is able to help them out of the place they are in to a better place. The Bible says "Ye are our letter, written in our hearts, known and read by all men" (II Corinthians 3:2).

I want you to think of yourself as a billboard. Someone purchases space to advertise whatever they are selling. Your life is a billboard in this world. Whether you are living for God or not—your life is on display. What will it say on yours? You are bought with a price. Jesus shed His blood on Calvary to pay for your sins and if you live for Him your life is a message to everyone you meet, to anyone you talk to, anyone who sees you. They may never step in church. But they see Jesus in you. How's He looking?

They may never cross the path of the church doorway, but they come across your path. They may never watch that TV evangelist, but they are watching you. They may never read one chapter in the Bible, but they are reading the story you tell every day by your words and actions.

Today, I want to encourage you to be Jesus to someone. Be Jesus' eyes. Look for the hurting and lonely and share a smile, or a kind word, or compliment, or a little hope. Be His hands. Reach for someone that's down and give them a hand up. Be His heart and love a lost and dying world. They are looking for Jesus. Will they see Him in you?

A Beautiful Soul

A beautiful soul passed from this earth not long ago. She was in honesty, the sweetest and kindest person I have ever had the privilege to know. She was beautiful inside and out. As I looked at her body lying in the casket, I realized that something vibrant and alive truly was gone. I have stood at many caskets in my life, but for some reason this one made me realize that these houses of clay that we walk around in are not really who we are. They are only our temporary homes.

We come into this world a living soul. The very God of the universe breathed life into each and every one of us. "The Spirit of God hath made me, and the breath of the Almighty hath given me life" (Job 33:4 KJV). I had that revelation the other day. I thought of how at every birth God blows that first breath into that new life and their soul begins to live. I truly believe He does that. We aren't fully alive until He breathes our soul into us. How beautiful is that? A living soul's birth. We are given a chance to live. We are put into this body of flesh that can experience all the wonders that a soul alone couldn't.

I can touch and be touched. My soul knows how love feels. When I run and feel the earth moving beneath my feet, my soul remembers what it is like to be free from the physical restraints that hold it in this body. When I sing, my voice lifts my soul to a place it remembers and wants to be—in a place closer to God. When I laugh or cry, I allow my soul to express in the physical what it couldn't on its own. This body is not really who I am. It is just a traveling suit. But it gives me a chance to live a life.

And today I am so humbly thankful for that. For all that I have experienced. The good and the bad, for it has shaped me into who I am. It has given my soul a chance to live a life and experience everything it could desire. I have always believed every single day we live is precious and shouldn't be wasted.

I am finding more truth in that every day that I live. At times this flesh fails, but that too, is all part of the soul's journey. Even in the darkest times, there is hope.

So smile. Laugh. Love. You have been given a life. A life with a purpose that God predestined just for you long ago, before He breathed that first breath into your soul. I pray that you find it. I pray that you live it. I pray that when your time is over, and they speak of your soul, they see beautiful.

MAKING MY MARK

I was so excited to be starting my new job. It was a brand new city. A fresh start. This was a new adventure, but also a life-changing one. This had been a big decision. It was the hardest decision I have ever made, and yet it happened very fast. Some things in life happen that way. Sometimes things seem to take forever, and other times God opens doors and you are ushered through in record time. Everything in my life changed.

If you have ever taken that big step to a new beginning, then you know exactly what I am talking about. Change can be scary. Uncomfortable. It's easier to stay where you are. It's safe and comfortable and familiar. That's what the place I left was like. That's not a bad thing. Hat feeling of security can be a wonderful thing. But it may not be the best thing. You can't expect different results when you keep doing the same thing. They say that is insanity. Amazing things can happen when you step out of your comfort zone.

So I was taking a gigantic leap out of mine. Go big or go home. The only problem was I had left my home far behind. So I needed to make this work. We all are on a journey. This was another step in my journey. This was a clean slate. It was an empty page where I was about to begin writing this new chapter of my story.

But here and now, as I was ready to begin my first day at this job, I realized I felt a moment of obscurity. It was a moment when I felt insignificant. I was the little fish in the big pond. I was the new kid on the block. I wanted to bring something of value to my new life. I wanted to fit in, but stand out. Make my mark. I wanted someone to know I was here. That I mattered. So I took extra care picking my outfit that day. Curled my hair. As I walked down the sidewalk toward the school's front door, I flipped my hair—something I tend to do when nervous, and I realized I still had a pink foam curler in the back. Oops! I quickly pulled that baby out before anyone could see, and dropped it in my purse. Or so I thought.

At the end of the day as I was walking back to my car, there on the sidewalk for all of the world to see (notably for any fellow sidewalk travelers to see) was my pink curler. I started to pick it up, and then stopped. In my head I heard a voice that said, "Wait a minute. Don't pick it up." So I didn't. You may ask, why even acknowledge that I heard that order, let alone obey it?

Now some of you won't understand my logical or illogical reaction (depending on your perspective), but some will see the point in it. Yesterday was gone. The past was over. People I've known, places I've been, things I've achieved. They are all a part of yesterday. But here...it's all new. All fresh. All unmarked. That curler on the sidewalk said something to me. It said that I was here.

At first, I guess I was just amused that it was still there. I started looking for it every morning and every evening. Silly, but true. I grinned as I stepped past it. And that curler laid on that sidewalk for over 3 weeks. No one picked it up or moved it. When I passed that poor curler (that was looking more decrepit with each passing day) I felt better. It was wet and smashed and sometimes covered with mud or snow or a combination of both. You would think it was a Hollywood star with my name on it. Not a sponge roller that finally was washed away into oblivion, or, in reality, a ditch. It was a little piece of me. It was a reminder that I was here now. I survived another day here. For every day it was there I gained a little more confidence. I breathed a little easier. I walked a little taller. I felt a little peace.

One of my favorite movies is "It's A Wonderful Life." We all know the point of the movie is realizing we all affect each other's life. We affect others in big ways and small ways. We make a mark on the lives of those we are around. Someone's life would be different if you weren't a part of it. The world would be different if you weren't in it. We all want the world to be changed because we're in it. We want to make our mark. But we can't do that if we don't take that step. There is a world out there waiting for what you have to offer.

But it's easy to get lulled into accepting the comfortable. Who wants to be uncomfortable? Not me. Not by choice anyway. The funny this is, even though it's not pleasant, discomfort never killed anyone. It may cause you to face harsh realities, make hard decisions, and even reinvent yourself, but it won't kill you.

Comfort on the other hand can be deadly. Being comfortable can kill inspirations and dreams. Some things have to be broken and remade to become their best. Some things have to be torn apart and began again to come out right. When we get too comfortable, we start being satisfied with the ordinary and the good enough. That's a lot easier than facing the unpleasant or uncertainties or the painful. We forget that out there is where the miraculous and extraordinary happens. Out there is where dreams come true.

I don't know where or when the curler disappeared. After a while I began to find my place and forgot to look. It may still be there off of that sidewalk, under the grime by the curb on one side or mixed in with the newly sown grass seed and straw on the other side. But the next time I need a little confirmation to help assure me, I will go for something more mainstream than a message from a curler.

I think I will do what any respectable graffiti artist would do. I'll just paint "Melanie was here" on the sidewalk for me and for all the world to see.

"You were tired out by the length of your road, Yet you did not say, 'It is hopeless.' You found renewed strength, Therefore you did not faint" (Isaiah 57:10 NASB).

Memories Are A Treasure

When someone you love becomes a memory, that memory becomes a treasure. I remember the night I lost my dad. It wasn't expected. Not really. I believed in my heart he would always be there. Just a phone call or drive away. We know that none of us will live forever on this earth. I don't believe ignoring the thought that we could lose someone we love is denial, but maybe for us humans it's a form of survival. One lesson I learned is not to take any day for granted. We are each only given a few short years. You may live to be one hundred, but in the big scheme of things, it's just the start of eternity. How I had wished I had visited more, called more, and talked more. Times that we could have shared were lost forever.

The night he died I sat deep in thought and remembered things from my childhood that seemed so insignificant. They were things that were just a part of my life. You don't realize all the things that you do and take for granted. They are just part of our daily life.

Memories played through my mind like an endless slideshow and I grasped on to each one. They were part of my life yesterday. That night they became a part of my history. Things I hadn't remembered before, now I didn't want to forget.

It is funny how in those moments things come back that had been long forgotten. But the mind and memory are amazing things. I can remember riding in the car with my dad, coming home from my grandmother's house one evening. He started singing the old hymn "In the Garden." That was the only time I can ever remember hearing my dad sing. I couldn't have been very old yet I can still remember that night and even now, hear the deep treble voice resonating as he sang. I still find comfort in that.

Hold your loved ones close every day. Never think of time spent together, even doing nothing, as wasted. On a night like tonight, someone will be remembering those moments. And while they seemed to matter little at the time, some day to a hurting heart, they will become a treasure.

You are Here For a Reason

You are here for a reason. If there wasn't one, you wouldn't be. It's that simple. You are smart enough to know that you are going to face opposition. The enemy knows who to go after. You wouldn't be in his sights if he didn't know where you were going and what you were going to accomplish. If God gives the hardest battles to His strongest soldiers, then it's time to soldier up.

Don't quit now. You may have been fighting so long and are tired, but you need to pull all your strength together for this one. Any good soldier knows that the fiercest, most intense battle comes right before the breakthrough. It may not look like no end is in sight in the middle of the battle, but keep going. You can make it out if you don't stop fighting. You may come out wounded, but fight on. Wounds heal. The scars will remind you of what you came through and how strong you really are. So get up and move forward. Someone may be depending on you to clear the way, and win this fight. There is more at stake here than you realize or can see. Fight on.

"Fight the good fight of faith; take hold of the eternal life to which you were called, and you made the good confession in the presence of many witnesses" (Timothy 6:12 NASB).

What Can I Bring The King?

I have always wanted to meet royalty. Maybe you have to. Have tea with the Queen. Chat and shop with a Princess. Dance with a Prince. I've thought about all the grandeur that would surround my visit. The stuff of fairy tales, right?

But it's not just the stuff of fairy tales. I have actually met royalty. Not just any royalty, either. The King of the earth. The King of heaven. The King of glory. The King of kings. It doesn't get any higher than that.

I think it is protocol that you bring a gift to the King when you visit—something worthy of so great a monarch. But what? It would be great to come before Him decked out in our finest clothes, riding up in our Rolls, the smell of our millions lingering in the air. Confidently walking up to where He waits.

In reality, it doesn't happen like that at all. Oh, I have come to Him, but I have come falling at His feet. I present Him with what I have to offer. A broken heart. Broken dreams. Broken promises. Brokenness. Have you ever been there? What is that to bring to the King of kings? And yet He isn't offended by the broken gifts. In fact, He seems to welcome them. He is moved by the tears. He is touched by the pain. He understands every feeling. That's why I come and fall before His throne of grace. He is a kind King. He has mercy that will hold me. He has grace that will help me.

But He is a powerful King. He will take all the broken

things we bring Him. You bring Him a heart that is broken and trampled and torn and scattered all around. You bring Him dreams that are shattered. You bring Him tattered bits of promises made but not kept. He gathers it all in His hands. Seeing Him hold all the pieces somehow makes you feel better. It gives you hope again. It gives you a reason to believe again. Because you know that when something looks impossible to fix—that's when He does His best work.

He is King of so many things, and He is the King of the brokenhearted. I can't think of any better place to put a heart that is broken. In His hands and care. You can trust Him to not only fix it, but make it better than it was before. He's just that kind of King.

"Let us therefore come boldly unto the throne of grace, that we may obtain mercy, and find grace to help in time of need" (Hebrews 4:16 KJV).

My Life In Boxes

Have you ever seen your life filling a bunch of boxes? Sometimes it's for good reasons. It could be a big move, leaving home to start school, or getting married. Or for a bad reason—like getting a divorce or even a death. That's what I looked at yesterday. Two decades of my life in boxes. Most of the things I really wanted I had already taken, but then there were the fragments left that fell in the what should I do with this category. If I wasn't missing it up till now, I really must not need it anyway. But I had to do it. I sorted through books, pictures, knick-knacks, kitchen gadgets and all the random stuff we seem to accumulate without even trying. I couldn't take it all, and actually there wasn't a lot left I even needed, but when it's your past—it's kind of hard to pitch it. I know it's just things, but it was a lot of pieces of—me. At least "the me" that I used to be.

So I carefully picked, and debated, and rationalized with myself on the choices, and then picking up the boxes I had filled, I felt it was finally ok closing the door for the last time and leaving with my past stash. When I got home, I suddenly got the urge to reorganize my closet. No simple task—believe me. Two hours later, as I stood knee deep in shoes, and random clothes, and of course tutus I wondered what had possessed me to do it. I looked at the clothes I had pulled and pitched—things I hadn't worn in forever. There were clothes that didn't fit my body or the person I had become anymore. I needed to get rid of them. Let go. It was at that moment that I realized something. The things I had gone through had changed me. Physically, emotionally, and spiritually. I caught my reflection in the mirror across the room, and I realized I am not the person I used to be. And that's good.

The past is gone, but the past is an excellent teacher. That's not a bad thing. Because of the past, you can be better. You are stronger. You know who you are and where you are going. Or at least you know the direction you should be heading. If not, now is a good time to get on the right track or road. You may get sidetracked now and then. I do, too. I have made so many mistakes, and I'm sure I'll make a lot more. I have learned more from the things I got wrong than from the things I got right. I believed things I shouldn't have. Trusted people I knew better than to trust, *but*—yesterday is past. We may not want to relive it or even revisit it. But don't let it define you. I have fought hard to start new. There have been hard battles, and long nights, and heartbreaks. That's ok. There have been victories, and sunrises, and joy. His grace has picked me up more times than I can count. His mercy has covered me, even though I don't deserve one bit of it. I find myself in need of them both today. But today is the first day of the rest of my life...and I am not wasting one day on what happened yesterday. Life even with all its twists and turns, ups and downs, beginnings and endings...is too precious to waste on the what-ifs. Keep moving forward. That's how you get there. I believe that it will all work out like it should and when it should. I might be surprised. It could be today. And if it is, at least my closet is ready.

"They will perish, but you remain forever; they will wear out like old clothing. You will change them like a garment and discard them. But you are always the same; you will live forever" (Psalm 102:26-27 NLT).

"Forget the former things; do not dwell on the past. See, I am doing a new thing! Now it springs up: do you not perceive it? I am making a way in the wilderness and streams in the desert" (Isaiah 43: 18, 19 NIV).

Tunnel Vision

Have you ever seen those Little Orphan Annie comics where she has those weird big eyes? It looks like she has some vision problem—like maybe tunnel vision. I realized recently that I have been suffering from tunnel vision. In case you don't know what that is, it's when you can only see what is right in front of you and lose sight of everything else. There are lots of things that can cause tunnel vision. It can be a symptom of certain diseases. It can also be caused by alcohol and drugs. (None of those were my problem.)

It can be caused by excitement or pleasure, and even from distress or fear. Evidently it is the extreme emotions that can bring it out. Now we are getting somewhere. That makes sense. We can be going along fine, and then the bottom drops out. That happens in life from time to time. No warning. No heads up. You just find yourself on the cold hard ground. If you weren't paying attention before, that fall alone will wake you up. Once the dust settles (or shrapnel stops flying in some cases) you try to get back up on your feet. It may not be easy either, because sometimes the pain and hurt leaves you wounded and bleeding. But you can't stay down. It's not safe to just lie there, even if you feel that's all you want to do. Worse things happen if you give up. You finally summon strength to go on.

But sometimes when you do get up, your vision is off. You aren't seeing things in the right perspective. All you can see is the one thing that went wrong. You see the one thing that fell apart. You see the one thing that you can't make sense of. That is where I found myself. Not proud to admit it, but hey, we have established that I am undoubtedly human. In the past two weeks, almost every day great things have been happening in my life. God is opening doors to wonderful things. Some are actually miracles in the making. But, I wasn't seeing them. My mind was focused on one thing. There can be all kinds of amazing things happening around you, but if your vision is fixed in one spot, then you are missing the total picture. Today, I realized that was exactly what I was doing. Life is like a tapestry that is being woven thread by thread. Where and when you look at the tapestry determines what you see. If you look as it is first being woven, the design may not make a lot of sense. As more is added, it starts to actually resemble something. It may take a while for everything to come together as it should...to be woven into something that we can recognize as good and beautiful. But one day, we will see how all the little pieces fit together to make the tapestry of our life. You have to be careful not to look back either. The back side of a tapestry looks dark and messy. Our past can look like that, too. Some things may have to be pulled out, and started over to get it right. That takes time and work and patience. Anything worth having takes all those things, too. Don't miss what is God is doing. Don't lose sight of what is important. God knows the path we take. He knows our end from our beginning. He has great plans for us between those two places. I am going to open my eyes and look past what is right in front of me to the bigger picture. There are great things happening. We just have to open our eyes and heart to see and believe.

"Open my eyes that I may see wonderful things in your law" (Psalm 119:18 NIV).

Happiness Is...

Charlie Brown said, "Happiness is a warm puppy." Look at his face. Doesn't that make you happy? It makes me happy. I am thinking today of things that make me happy. Fields of flowers. Sunrises. Sunsets. The ocean. Belly laughs from kids. Holding hands. Baby animals. Big animals. Tutus. Running. Need I go on? I think you get the picture. There are so many things that make me happy. I have always been a happy person. Because life is so perfect? It is anything but that. But I have found the secret to being happy. Scientists now say there is a happiness factor. Interesting. It is 50% genetics, 40% attitude, and 10% circumstances. According to scientists that would mean it's mostly in your head. You can change the happy factor in your life by changing your mindset. You are the keeper of your happiness. So where are all the happy people? From my psych classes I learned that there are more people on antidepressants now than ever before. Is it because they have nothing to make them happy? The answer to that would be...no. What is being happy about anyway? The dictionary says it is the state of being content with your situation. Hmmm. I beg to differ. I think finding true happiness lies in being able to be content no matter *what* your situation. Some people think if you have enough money, you can buy happiness. (Wait...isn't there a curse of the millionaire?) It is true that you can buy about anything. Just look on eBay. Things aren't bad. There are things all around us that add fun and joy to our lives. That is so true. But it isn't really about the things. It is how you "feel" about the things. You can get all kinds of things, but if you aren't happy inside, you won't find the happy with them. I love to run. Running doesn't make me happy. It's how running makes me feel that makes me happy. I feel alive. I feel strong. Sometimes I feel tired. Sometimes I feel sweaty. But I always feel happy. One thing life has taught me is that life is more than what I possess. Another thing that life has taught me is that there is so much more to life than what is going on around me. I have learned to find my happy place in spite of what is going on in my life. When you realize life will never be perfect, it takes a lot of the pressure off. Does that mean you are never sad? Of course not. But it's in those times that you learn to look for a reason—any reason to be happy. At the darkest times you can always find one if you try. A tiny sprout must force its way up through the crushing weight of the soil to get to the light, and have a chance to live. Sometimes we have to find that tiny sprout of hope, and force with all we have to live. That is what I have learned to do. I look for the happy in everything around me. The morning sun. That first cup of tea. Being greeted by a favorite dog. The smell of cookies baking. Playground games with the wee ones. The sound of my running shoes hitting the ground. Talking to a friend. Eating tomatoes right out of the garden. That first tug on your fishing line. Getting mud soaked. (Yes, I like mud.) Things that seem so insignificant, but they're not. They make me happy. Finding the happy in the little things is what helps us get through the big things. On my darkest days, it is the little things that hold me together. I came home on Monday to a box of peacock eggs waiting for me to hatch. (I actually tripped running so fast up the steps to get to the box.) That is probably not something most people would put in their happy column, but it made my day. The eggs. Not the trip. No matter where you find yourself today or what you may be going through, look for a reason to smile. It doesn't matter what it is as long as it leaves you feeling—happy.

"I am not saying this because I am in need, for I have learned to be content whatever the circumstances. I know what it is to be in need, and I know what it is to have plenty. I have learned the secret of being content in any and every situation, whether well fed or hungry, whether living in plenty or in want" (Philippians 4:11-12 NIV).

"A happy heart makes the face cheerful, but heartache crushes the spirit" (Proverbs 15:13 NIV).

Waiting For Summer

It seems like it takes forever when you're waiting for something to come. The more you want it...the harder the wait. I remember waiting for summer to come. Last winter made it seem like it would never arrive. But as I sit on the deck with my bare feet basking in the sun, clear blue sky above me, and listening to the sound of water babbling over the rocks in the stream, I know summer has definitely arrived. You have to love summer days.

It seems like just the other day, I was covered from head to toe in cold weather gear, battling through one endless snowstorm after another. Remember Randy in The Christmas Story? That is how I look on the playground at school in the winter, too. But not today. Yesterday, it seemed like winter would never end. Just when you think you can't shovel one more snow-covered sidewalk or dig your car out from another snow burial, something begins to change.

One day you begin to notice the air is not quite as brisk as it was. The sun actually feels warm. You can see the grass, even though it is brown and smashed from the weight of the snow that has covered it mercilessly. Seasons end. We know they eventually have to. That is just how the seasons work.

Life is about seasons, too. They are always slowly changing—just like the seasons in the world around us. Winters in our life are tough. They take us through cold, frigid, dark places where the sun doesn't shine.

They leave us shivering and yearning for days that will make us feel warm and alive again. I have been through those winters. Maybe you have, too. You don't think the sun will ever shine again. You desperately want the pain and loneliness to end. You want new life and new things to begin in your life. But all you see is winter. But nothing lasts forever. Not the good or the bad. Not even the winters of our lives. Change will come again.

The coming of spring is almost a magical experience. One day little leaves begin budding on the trees. Flowers burst out of the softening ground. The grass actually starts to look its rightful color. Green. I don't know if it is the relief from the cold or that we start getting more active that it suddenly seems overnight the world is green and growing and alive. We were waiting for spring. But suddenly summer is here.

We experience the same phenomenon in life. All can be dark

in our life, then suddenly things start to change. We begin to feel the weight of what we have carried through winter lighten just like we trade our heavy coats and boots for shorts and flip flops. We are not sure how it happened. We had waited so long to get to enjoy life again and now we are. Nothing lasts forever. Not the pain—or winter.

I am enjoying every sunny day, every blue sky, every warm night. I will enjoy every glorious moment this summer. Because just like winter doesn't last forever... neither does summer

"But seek ye first the kingdom of God, and his righteousness; and all these things shall be added unto you" (Matthew 6:33 KJV).

"As long as the earth remains, there will be planting and harvest, cold and heat, summer and winter, day and night" (Genesis 8:22 NLT).

First Snow

Snow is beautiful. It has a magical quality to it. There is something about waking up to watch the falling snow covering the ground and trees and mountains in the distance that makes the world seem softer and kinder. It's so peaceful and quiet. In this day that we live, I can truly appreciate anything that can make you feel that way.

I think snow looks so beautiful before anyone steps in it. As I left for work yesterday, it wasn't quite daylight, and the snow glistened like a million diamonds lying in my path. I felt bad walking through it, but without my fairy wings, I had no other option. I hate to mess up beautiful things.

Do you ever think about how life can make things so ugly? Things can get really messed up. I don't want to see the ugly in life. I want to see the beauty. But it doesn't always work out that way.

You have this beautiful snow-covered life. That in itself would be wonderful—if it stayed like that. But, it never does. Sooner or later, someone is going to step in your snow. They are going to leave a dirty footprint here and a boot print there. At times it feels like they are scooping away parts of you and tossing you like snowballs in a fierce snowball battle. You know when they are done with you, you will be a splat on the ground. Or smashed on the wall of life.

Before long the pristine veneer is gone, and what's left? Slush. Slush is really the opposite of that beautiful, unspoiled snow. Some slush is not really bad. It still looks white, but it's melting. Not so perfect and hard to hold together. I have been in the slush stage before. Things around causing me to fall apart. Things get hot and I feel like I am melting away. It's hard to keep yourself together when you're falling apart.

Then there's the dirty, ugly slush, melting and mixing with dirt and mud. Not pretty. I have been that slush, too. I have let things into my life that spoiled the white and pure beauty of my snow. Made me dirty. Made me weaker. Made me ugly.

You know I am a firm believer in always looking for the positive, no matter how small, in everything. The belief that everything will work out. That everything will come together. If we lose that hope, we lose everything. So where do you go when you find yourself turned into a slushy mess?

What hope is there for finding a way to change the muck and mire of your slush into anything of worth? Sadly, you can't. Try as you might, you can never make yourself good enough or clean enough on your own.

But as I stand and look through the glass door to the untouched snow covering my world, I remember. "Come now, let us settle the matter, says the LORD, Though your sins be as scarlet, they shall be white as snow; though they be red as crimson, they shall be like wool." Have you ever seen the wool of a new lamb? One of the purest things I have ever seen is a newborn lamb. They are so soft and white. Not blemished by the world and dirt around it yet.

My "snowy" life at times has been marred and scarred by the pain I have endured and mistakes I have made. I was left a slushy pile waiting to dissolve into oblivion...or the ground...whichever came first. But even at my filthiest moment, there was hope. A hope that like that untouched snow...covered me, cleansed me, and made me beautiful again. The dark crimson stains that made me feel so guilty were washed away and like the magical feeling that comes watching the first snow fall, I found a peace to start again. If you fall, don't stay down. We all fall sometimes. If you fall along the way, it's not as bad if someone is there to help you see all is not lost ...other than your dignity and footing. They can help you laugh, or cry, and brush the "snow" off and keep moving in the right direction. Snow can be slippery and so can life. It can be treacherous stepping, but keep going. You'll find your footing. When you do, you may need to help someone else. If we hold each other up when this life is unstable, we have a better chance of making it where we're going.

"Purify me from my sins, and I will be clean; wash me, and I will be whiter than snow" (Psalm 51:7 NLT).

"Two people are better off than one, for they can help each other succeed. If one person falls, the other can reach out and help. But someone who falls alone is in real trouble" (Ecclesiastes 4:9-10 NLT).

Silent Night

"Silent night. Holy night. All is calm. All is bright..." And all is well. You know what I love about Christmas? With my love of shoes and tutus and pretty clothes...you probably think it's presents and shopping and sales. You couldn't be further from the truth. If you have realized much about me, you know I love to put together styling outfits (but I like to do my shopping at thrift shops—that's a whole other story).

Christmas has never been about presents for me. I honestly don't care if I get a single one, and God knows that is the honest truth. Christmas has always been about the feeling that comes when December rolls around. It starts the day after Thanksgiving. And no, it's not Black Friday shopping. I shouldn't admit it, but I don't Black Friday shop. I have only gone once in my life. They may revoke my woman license now, but I don't care.

I don't start Christmas planning or activities or decorating until the day after Thanksgiving. I am old fashioned, I guess, but life is so rushed and hectic. I just need some things to stay constant. On the day *after* Thanksgiving the countless boxes, and branches, and wreaths are drug out and carried up and strewn across the living room. Then comes the monumental task of unloading and arranging everything.

It sometimes takes me two days to decorate the tree. No joke. Of course, I like to put an ornament on *every* branch. I like to savor the job, too. Because the tree isn't just a decoration, it's a gallery of memories. As I decorate, I relive and revisit times and memories over and over again. Most of my ornaments are homemade. Every year I add new ones. Some are themed on events that have happened in the year.

Others may be something special we liked that year. I have every conceivable Christmas icon made from little hand prints and footprints, along with ornaments made from everything from pop cans to egg cartons. Then there are the ornaments that were gifts from others. I have ones from loved ones long gone, yet every year as I place their ornament on the tree, I remember them, and I feel them near. It may sound gaudy, but the tree turns out beautiful every year. Or maybe I just feel it does.

But see, that's what I am talking about. I love the feeling that comes with Christmas. It isn't imaginary either. Everyone is just a bit nicer and kinder. Even people that we dub as Scrooges and Grinch's through the year seem to mellow as the holiday nears. There is peace on earth, or at least in our little town. You can't find that at a Black Friday sale, or even in a present under a tree.

It came one Christmas many years ago, wrapped yes, but not in pretty paper and bows...no, it came wrapped in a soft, blanket. Not to a lavishly decorated home or palace or even... a hospital. No, He came into a cold stable. No colorful strands of Christmas lights...just one bright star. Not to the sound of cantatas filling an auditorium...but to the angel choir's announcement that peace on earth and goodwill to men had come.

And it still comes. Every December, especially as we get nearer to the 25th. But it doesn't have to be allotted to that time. No, His peace can be a part of everyday—all year long. If only that really happened. How different the world would be. How different my life and your life could be. "Peace I leave with you; My peace I give unto you; not as the world gives do I give unto you. Do not let your heart be troubled, neither let it be afraid" (John 14:27). His peace lasts, not just for one month, but for a lifetime.

His peace will carry you through. No matter what you face or what may come...let His peace surround you today, Christmas day, and every other day of the year. All is well.

Emmanuel

The angel told Mary to call His name Jesus. The angel told Joseph to call His name Jesus. The prophet of old said He would be called "Emmanuel." God with us. Our minds turn toward that night over 2000 years ago when He came to a dark and hurting world. He was born to a poor, young couple from a small town. Not at a palace full of riches and servants. He was the

King, but that wasn't the plan.

Emmanuel, God with us. He grew up working alongside Joseph, learning carpentry, which was no wimpy trade in those days. He stood by His mother as a teenager when they buried Joseph. He had to help carry the load His mother bore as a single mother then. He wasn't a stranger to poverty, or death, or having to work hard. He wasn't a stranger to the struggles and hardships of life. He knew rejection and how it felt to be an outcast. He was human, just like us. He felt the pain, the temptations, and the heartaches. Just like us. He was born and raised among people just like us. People that were hurting and struggling—just like us.

The world is still dark and hurting.

Every year we set up the nativity. We put the baby in the manger. We place all the figures of the story in place. After Christmas we pack it up until next year. But you know what? I need Emmanuel. Not just at Christmas. I need Him every day. I can't put the baby back in the box. No. I need to keep Him out. I need the promise that the angel gave.

Emmanuel. God with us. He came that dark night so long ago to bring hope. I need that hope now. Just like His mother wrapped her arms around her newborn Son and held Him near, I need Him to wrap His arms around me and keep me safe from the doubts, fears, and pain that try to overtake me. The angels sang about peace on earth. I need that peace to fill me like the angel choir song filled the skies over Bethlehem that night.

I need Emmanuel. God with us. It's not a promise just for Christmas. It's for every day. I need Him. He is here to be Emmanuel to us. When we are struggling. When we face temptations. When we are hurting. He doesn't wait for us to put on fancy clothes and proper manners to present ourselves to Him. No. He loves us just like we are. He knows how it is and what life is like and what it means to be human. He wants us to remember that the promise the angel gave long ago on that night He came to the world still rings true. He has come to be Emmanuel. God with us. Then and now. You are never alone.

"Behold, a virgin shall be with child, and shall bring forth a son, and they shall call his name Emmanuel, which being interpreted is, God with us" (Matthew 1:23 KJV).

The Journey Continues

Have you ever played the game of Life? Oh, I know we feel like we are all playing that game every day. I remember spending many hours around that colorful board, deciding which roads to take and facing the choices those decisions brought. College or not. Get married or not. Have kids or not. Kind of heavy for a kids' game, but now looking back, it was one of my favorites.

This life always poses the same the question that the game did...where do we go from here? That is a question that we ask ourselves many, many times throughout our life. Where do I go from here? Wouldn't it be great if when you were born, you came with a map showing exactly the way you should go to get to where God wanted you to be? It would be all laid out right there before you. You would never wonder if you were headed the right way. Not sure? Check your map. How far do you need to go? Check your map. Should you take this shortcut up ahead? Check your map. Unfortunately, life doesn't come with one. It would be so much easier.

Life is filled with different roads that take us different places. We want to follow the ones that lead us to our destiny, the destiny we were born to fulfill. The only problem is being able to know which way to go. How can we be sure we are headed the right way? That is where faith steps in. "We walk by faith, not by sight" (II Corinthians 5:7). It is really hard to walk without seeing.

Ever played Blind Man's Bluff? You close your eyes and the next thing you know... you run into a tree. That has happened to me in life. Not every road we travel is clearly marked or even well lit. There have been times I couldn't see where I was going, and the next thing I knew, I was face down on the ground.

It is in those times that I realized I took a wrong turn. But even when we get off course, we can find our way back. No matter how far you have gone down a wrong road, you can always turn around. It's not the end of the journey, just a momentary detour. We may have to cross some rough roads to get back to where we should be, but we're not alone. He promised never to leave us or forsake us, and I have found Him to be true to His word.

Even when I have messed up badly, He has never once turned me away. No, He has picked me up, wiped my tears, and helped brush away the bits and pieces of the mess that I fell into. Some things have been painful. Some have been devastating. Some have left deep scars. Those scars remind us that we survived. It may have broken us for a while, but it didn't kill us. As long as we keep going, we recover, heal, and find our way back.

Walking by faith can feel as unnatural as walking without looking. But just like walking with your eyes closed, the more you do it, the more comfortable you become. You learn to have a sense of not only your surroundings, but the direction you should be going.

Even with all the ups and downs, life is an adventure meant to be lived, not just followed on a map or game board. We become who we are by the things we experience, the people we meet, the choices we make. Life is meant to be lived.

I have grown and changed more in the past few years than in any other time in my life. I have seen my worst days, and my best days. I have endured my saddest days and happiest days. It has all been part of my journey. Have I always followed the exact road I should? No. Who does? But I have kept going. I may lose my way from time to time, but that doesn't change the destiny He has for me, as long as I keep moving ahead.

It is everything that we face and experience and live through that makes us the person we are. When we get to the end of this road, we want to end up where we supposed to be. I may end up looking like a contestant from Survivor, but I want to finish this journey strong. I may be tired. I may be sweaty and dirty. My clothes may be torn from the tangles that I have had to fight through along the way. I may have to crawl. But, no matter what it takes, I want to make it where I am going. I want to make it where I need to be.

This journey goes on for now. We never know how far we have to travel. But, remember we are not traveling alone. We are walking this journey out together. I pray that you will keep going, no matter what comes your way. I pray you find love that never fails, peace that passes all understanding, and strength to make it through everything you must face. Even when it seems you are alone—look around. We are all traveling this road together. You are never alone. Reach out to someone. Take their hand and help them walk. Or let them help you. We need each other. That is part of the joy in the journey. We were never meant to travel alone. We are all just walking each other home.

Made in the USA
Middletown, DE
27 June 2023